MATHEMATICS
Coaching Handbook

Working with Teachers to Improve Instruction

Pia M. Hansen

EYE ON EDUCATION
6 DEPOT WAYWEST, SUITE 106
LARCHMONT, NY 10538
(914) 833–0551
(914) 833–0761 fax
www.eyeoneducation.com

For information about permission to reproduce selections from this book, write: Eye On Education, Permissions Dept., Suite 106, 6 Depot Way West, Larchmont, NY 10538.

Library of Congress Cataloging-in-Publication Data

Hansen, Pia M., 1955-
   Mathematics coaching handbook : working with teachers to improve instruction / Pia Hansen.
      p. cm.
   Includes bibliographical references and index.
   ISBN 978-1-59667-093-8
  1. Mathematics teachers--Training of--United States. 2. Teachers' assistants—Training of--United States. 3. Teachers' assistants—United States. 4. Mathematics—Study and teaching (Elementary)—United States. 5. Mathematics—Study and teaching (Middle school)—United States. I. Title.
   QA10.5.P69 2008
   510.71--dc22

                                        2008026424

10 9 8 7 6 5 4 3 2

Production services provided by
Rick Soldin, Electronic Publishing Services, Inc.
Jonesborough, TN — www.epsinc-tn.com

Also available from Eye On Education

Meet the Author

Pia M. Hansen has been a classroom teacher for twenty-seven years, teaching students from pre-school to college level. Currently, she coaches teachers engaged in modeling lessons, teaming, teacher observation, examining student work, and collaborating on district and school wide assessment practices.

She is the co-author of *Performance Tasks and Rubrics: Primary School Mathematics* with Charlotte Danielson and *Buddies,* a cross grade mentoring approach to implementing standards, both published by Eye on Education; and the third grade curriculum *Bridges in Mathematics* published by the Math Learning Center.

Contents

Preface

I loved children and wanted to make a difference; therefore, I entered my teaching career thinking I was ready for the daily challenges! I soon recognized how ill-prepared I really was. Situations would arise that caused me to question how to do this work called education. My students, their families, and my adult colleagues continued to enhance my journey. Thirty years later, I am so grateful for the "coaches" in my teaching career. Some of my coaches were master teachers that opened their classrooms to me after school. They questioned me, they listened, and somewhere between their classrooms and mine, I found my answers. Some of them were colleagues who took classes alongside of me and shared my interest in learning. They tried new ideas and were willing to candidly discuss what worked and what didn't. Some teachers challenged my assumptions and my direction, and in doing so, made me accountable for data. Some of these colleagues were student teachers who questioned my every move and pressed me to explain why and how and what next. Some of them were children who knew best what they needed most. Some of them were parents and family members who knew their children in a way I hadn't considered. Some of them were principals and editors who believed in me and thought I could be a coach today.

I was a Montessori pre-school teacher and administrator of a school for ten years. I taught country school and city school in Wyoming for 19 years. I taught courses for the university and community colleges, and presented at NCTM and ASCD conferences along with some local venues. I had an opportunity to publish books on *Performance Tasks and Rubrics for Primary Grade Mathematics*, with Charlotte Danielson; *Buddies* (Cross-grade mentoring); and *Bridges in Mathematics*, a comprehensive third grade mathematics curriculum. Most of the examples in the manuscript are lessons from this exceptional curriculum. I was a standards specialist on the cusp of standards-based teaching and learning, working alongside teachers in their classrooms with their students. Then, another door opened. I was asked to provide professional development workshops on standards, assessment and teaching practices, especially in the area of mathematics. I knew the implementation of any new ideas depended on follow-up with classroom modeling and observation visits. Many districts and principals found the time and money in their budgets to allow this to happen on a regular basis. Their implementation of the professional development increased student achievement. Other schools struggled to impact practice and change beliefs. I studied both scenes to determine how to make a bigger difference.

Coaching is a professional and personal journey. It requires a genuine passion for working with children and their teachers. It requires a thirst for learning, reading about the research, and then implementing best practice. It also requires taking risks, jumping in to teach a lesson, teaching colleagues, and navigating some tough political waters, too. I am indebted to the profession I serve. How lucky I am to have found my way!

I dedicate this book to my children, who taught me about coaching through the heart and eyes of a parent: standing on the sidelines, cheering at soccer games, watching you ride horses, and holding my breath at triathlons. I couldn't make the goal, turn the horse around, or run the race for you, but I learned how to help you find your way. I also dedicate my work to my mom — my first teacher and coach. She and I joke about how someone who was never an athlete or at the top of her math class could be a math content coach now!

1
Examining the Role of a Math Content Coach

"Faith is taking the first step even when you don't see the whole staircase."
Martin Luther King, Jr.

Current practice in mathematics teaching has been impacted by the National Council of Teachers of Mathematics Content and Process Standards (NCTM 2000, Appendix 1), NCTM Curriculum Focal Points for K-8 Mathematics (NCTM 2006, Appendix 2), and the Federal No Child Left Behind Act (2001). As states, districts, principals, and teachers consider how to increase student achievement in communication, problem solving, and complex reasoning strategies, content coaching initiatives for mathematics have begun to emerge. In some cases, teachers have left the classroom to fulfill this call to action with very little support and preparation. This book investigates the fundamental nature of teacher leadership and the essence of content coaching in mathematics.

Having faith in the coaching process requires us to take the first step: understanding and planning. When we begin to walk up a staircase, we don't see every step but we need to trust in the process; it will take us to the landing. This first chapter defines and examines the role and responsibilities of a building math coach as a school leader and the importance of developing professional learning communities. Given that each school and district initiative is unique, administrators and math coaches will need to determine what fits their specific site.

Defining the "Math Content Coach" Position

Math content coaches provide professional development in mathematics content and instructional pedagogy. Initially, they might help schools examine their resources for teaching mathematics: *How much time is being spent teaching math during the school day? Are there funds available for purchasing math curriculum, supplementary materials, and required manipulatives? Are additional people available to help teach math in smaller groups, support classroom teachers in preparing materials, tutoring at-risk students, and challenging high potential students?* Coaches, along with building leadership, may suggest ways in which to allocate these resources more effectively. Throughout the school year, coaches provide ongoing professional development with an eye on increasing student achievement in mathematics in a variety of ways. Each site will determine the math coaches' roles and responsibilities based on their needs. Below are a few suggestions to consider.

Roles and Responsibilities of Math Content Coaches

♦ Support the professional growth of elementary and/or middle school mathematics teachers by increasing classroom teachers' understanding of math content

♦ Enhance mathematics instruction and student learning by helping teachers develop more effective teaching practices that allow all students to reach high standards

♦ Collaborate with individual teachers and teacher teams on planning standards-based units, modeling, team-teaching, and coaching cycles

♦ Share research about how students best learn mathematics and facilitate teachers' use of these instructional strategies, including differentiated instruction for diverse learners, manipulatives, and visual models for mathematical thinking and reasoning

♦ Facilitate opportunities for teachers to develop an understanding of the national, state and district math standards and grade level benchmarks and to identify the "essential learning" in mathematics for their students

♦ Assist administrative and instructional staff in interpreting student achievement data and designing approaches to improve instruction and student learning

♦ Examine classroom-based math assessments and standardized test items for alignment, cognitive demand, equity, and purpose and use this information to make instructional decisions

♦ Examine student work as evidence of understanding, misunderstanding, and proficiency and use this information to make instructional decisions

♦ Provide staff development for the math curriculum or adopted math supplements

♦ Organize professional math resources such as reading and teaching materials

In addition to *content coaches* that work with all teachers, some districts have also hired *mentor* teachers. Mentors are usually experienced teachers who are assigned new-to-the-profession (or new-to-the-district) teachers with the goal of initiating them through a process of induction and support. These mentors are generally responsible for several content areas, enhancing classroom and materials management skills, dealing with logistics, and setting up curriculum, instruction, and assessments schedules. Many of these responsibilities may overlap with the roles of math content coaches. In September of 2007, NCTM released a position statement on mentoring new teachers:

> *. . . These partnerships should ensure a strong focus on mathematics content knowledge, pedagogical knowledge, and knowledge of Principles and Standards for School Mathematics (NCTM 2000) and its application to the classroom.*

Perhaps the two most significant differences are that math content coaches focus on mathematics teaching and learning and include veteran and novice teachers. Mentors may not be able to support a comprehensive understanding of math content and pedagogy knowledge or have the time to work with veteran teachers. Both coaches and mentors need to be consistent, flexible, respectful, willing to engage in dynamic learning, and committed to increasing student achievement. Their interactions are built on fairness, confidentiality, privileged communication and therefore should not be used in a formal evaluation process. Coaches and mentors are NOT administrators, supervisors, or evaluators of teachers.

That said, often principals rely on coaches and mentors for "another pair of eyes" on classroom situations: *How's it going? How much support does a teacher require to effectively teach math? What other resources need to be allocated to help the teacher be successful?* One size doesn't fit all, and therefore coaches and mentors make professional recommendations and decisions based on what's in the best interest of the students. When they recognize a critical situation, they act in a professional manner.

> *At a grade level follow-up meeting in November, I discovered a teacher was a whole unit behind in her instruction because she had not allocated enough time for teaching mathematics, did not like teaching math, and claimed her students didn't like the new program. This was going to significantly impact student achievement, and therefore it was my responsibility to meet with the principal and inform him of the circumstances. We collaborated on a plan that would include organizing her classroom schedule and materials and providing two weeks of in-class modeling on how to teach this new standard-based program. We also planned one week of monitored guided practice with specific feedback on key instructional strategies and a biweekly meeting to ensure a gradual release of responsibility. This intervention would not have worked without the principal's participation and a professional coaching relationship with the teacher. By February, the teacher was on target with her instruction and surprised at how much her students were learning!*
>
> *—Math coach reflection*

Coaches as Leaders in a Professional Learning Community

Math content coaches relate as peers and focus on reflection, content, and teaching and assessment practices. This specialized inquiry focused on student learning in mathematics creates a true professional learning community, sometimes beginning with just a few teachers and coaches. DuFour & Eaker (1998) define the Characteristics of Professional Communities:

Characteristics of Professional Communities

♦ Shared mission, vision, and values

♦ Collective inquiry

♦ Collaborative teams

♦ Action orientation and experimentation

♦ Continuous improvement

♦ Results orientation

Coaches are leaders in the establishment of these communities. They commit to guiding principles: the mission, vision, and values that identify what the school believes and what they seek to create. Math content coaches also commit to the implementation of a standards-based mathematics curriculum, via the process standards, and authentic assessment practices. They commit to increasing their own and teacher content and pedagogy knowledge. They question the status quo, seek new teaching methods, test those methods, and then reflect on the results. Coaches that make a difference don't just go along with what's been happening in the building. Their professional curiosity and openness to possibilities inspire team members to develop new skills and take risks as well. These experiences lead to genuine shifts in attitudes and beliefs about student learning.

Math coaches increase building capacity because they recognize that organizational growth depends on collaborative teams, not independent work. Teaching can be an isolated profession, where each person has a room with a door that remains closed during the day. The students in the classroom are the responsibility of that one teacher. In some schools, teachers are actually encouraged to compete with each other for parent requests or higher test scores. Coaches believe that teaching should not be an isolated or competitive profession and seek to open those doors in a figurative sense. They encourage teams of teachers to take ownership of all their students, share their best practice, and promote inquiry about their profession. Many schools already have pockets of professionals that work collaboratively. Content coaches strive to make these exceptional relationships intentional and expand them to include all the teachers in the school. If students need to be a community of learners in the classroom, then surely teachers need to develop as a community of learners, too. Effective coaches make connections, one-by-one, and in small groups, towards the development of a schoolwide community that learns together. This intradependence creates the synergy, the momentum for positive change and renewal.

Coaches take action. They may order new math curriculum materials to examine, organize storerooms full of math manipulatives so that teachers can find what they need to increase student engagement and understanding in their lessons, or look for math conferences and classes that align with best practice. They DO something to get the ball rolling! Math coaches take initiative.

They understand that engagement and experience are the most effective teachers. So, they model math lessons; observe, collect, and examine school data; read and reflect; develop, test, and evaluate learning theories. Coaches with intermediate teaching experience might be asked to coach in a K-6 elementary school. They know they need to experience primary grade level mathematics as a learner and as a teacher, so they may schedule some observations in a first grade room, observe classroom management skills and routines, and then jump in to teach a series of math lessons. Coaches with elementary teaching experience may do the same in secondary classrooms. In order to establish and maintain credibility with their fellow teachers, they must be able to walk the talk.

Coaches also understand that their work is never done. They celebrate significant milestones and chart the course for the next journey. They appreciate that being an educator is a lifelong pursuit.

School, Classroom, and Student Factors

Schools that have principals with strong leadership skills and effective content coaches are likely to succeed in their mission, vision, and values. Content coaches alone will not be "the magic bullet" to increasing student math achievement.

According to Marzano's research on *School Leadership That Works* (2005), there are eleven factors that create effective schools. Below are eight factors that are, at least partially, the responsibility of the math content coach. The first three are identified as school-level factors, the second three are teacher-level factors, and the last two are student-level factors. School-level factors are based on policy and include schoolwide initiatives or operating procedures. Teacher-level factors include instructional strategies, classroom management, and curriculum design. Most of the math content coaches' time will be spent in this area. Student-level factors include the home atmosphere and influence.

School Factors

- ♦ **A guaranteed and viable curriculum.** Ideally the curriculum in the school is aligned to the NCTM Content and Process Standards, NCTM Focal Points and the State Standards. Once this has been established, it is imperative that there is enough instructional time allocated to teach it (viability) and that every teacher is teaching the essential content (guaranteed). In many schools, the literacy block is sacred, but the math block is frequently interrupted by school assemblies, field trips, guest speakers, early release days, and so on. By the end of the year, students may have missed almost 17 hours of math instruction. Building leadership and math coaches can help teachers schedule and protect the instructional time needed for mathematics. Coaches can help teachers identify and communicate the essential math content for all students (what do they need to know and be able to do), and with the support of building leadership, ensure that this content is addressed in every classroom in the building.

- ♦ **Challenging Goals and Effective Feedback.** The second factor requires implementing an assessment and record-keeping system that provides timely feedback and establishes and monitors rigorous achievement goals for the school and individual students. In many districts, report cards (electronic or paper) are created at the district level. This reporting system may or may not be adequate and may need to be reviewed by the content coaches. In effective schools, assessment drives instruction, and therefore coaches ensure the math assessment data is high quality and communicates specific feedback on academic progress to the students, parents, and teachers.

- ♦ **Collegiality and Professionalism.** Along with building leadership, coaches help establish norms of professional conduct and behavior that engender collaboration, encourage teachers to participate in school decisions and policies, and provide teachers with meaningful staff development in mathematics that is focused and cohesive. This factor mirrors the work by DuFour & Eaker on Professional Learning Communities.

As schools move toward "teaching to the standards," classrooms take on a new look. Rows of isolated desks are rearranged in pods that encourage students to share their learning materials and communicate their ideas in pairs and small groups. Student access to an overhead, document camera, or interactive white board encourages student representation of mathematical ideas and promotes rich discussion of their reasoning and proof.

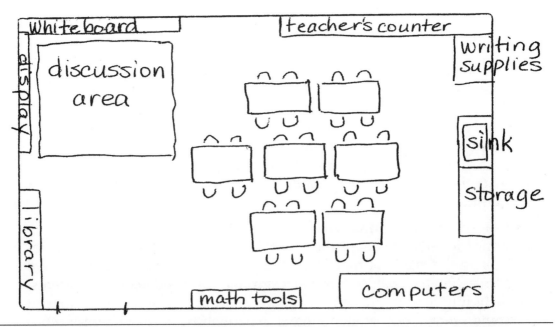

This classroom layout encourages student to student interaction and participation.

More than just a physical setting with desks, bulletin boards, and posters, the classroom environment communicates subtle messages about what is valued in learning and doing mathematics.

NCTM, 2000

Classroom Factors

♦ **Instructional Strategies.** Effective teachers and coaches have a variety of instructional strategies at their disposal. They use these strategies across all content areas, not just for math. These nine strategies are *identifying similarities and differences, summarizing and note taking, reinforcing effort and providing recognition, homework and practice, nonlinguistic representations, cooperative learning, setting objectives and providing feedback, generating and testing hypotheses, cues, questions, and advanced organizers.* Veteran and novice teachers alike benefit from an explicit review of these strategies. Math coaches make these strategies explicit during the pre-conference, modeling of the mathematics lesson, and post-conference session, and part of their professional development workshops. These strategies are often the focus of a lesson/unit study approach in a coaching cycle.

♦ **Classroom Management.** Teachers articulate and enforce rules and procedures, develop a respectful relationship with students, employ specific strategies to increase their awareness of special needs in the class, and maintain a healthy emotional objectivity with their students. When coaches go into a classroom to model a math lesson, they are aware of the rules, procedures, and consequences and make clear their expectations for behavior. In the pre-conference specific student needs are identified so coaches can be responsive to them during the lesson. In some classrooms, math coaches may need to model how to manage a variety of manipulatives during a lesson and employ specific cooperative learning strategies, like a three-minute review (Teachers stop any time during a discussion and give students three minutes to review what has been said, ask clarifying questions or answer questions).

♦ **Classroom Curriculum Design.** Teachers and coaches make instructional decisions to adapt the content found in math curriculum materials to the needs of their particular students. These adaptations are focused on the standards and background knowledge of the students, including how to present this information in a way that is similar to some other topics students have experienced, perhaps across other content areas. Teachers and coaches consider which math skills are introductory and which ones are taught for mastery, and how to challenge students with complex tasks that require them to apply their new knowledge to generate a generalization or hypotheses. Many of these decisions are made during a pre-conference planning session. Some decisions are made "on the fly" during the lesson itself. These curriculum decisions are re-examined during the post-conference.

Home Factors

♦ **Learned Intelligence and Background Knowledge.** Life experiences certainly contribute to student academic success, and therefore math content coaches work with colleagues to engage students in a variety of simulations, reading, and vocabulary activities that are linked to mathematics teaching and learning. These tasks are systematically linked to the assessment practices that measure student achievement. Math coaches and teachers communicate these expectations to the families of the students they serve.

♦ **Motivation.** Coaches work with teachers to identify worthwhile and rigorous mathematical tasks, with multiple entry points, to enhance student motivation and engagement. These problem-based projects inspire even reluctant learners to use manipulatives, models, and real-world contexts. Students make choices, invest their time and resources and thereby increase their motivation.

Home Connections For use after Unit Two, Session 7.

NAME _____ DATE _____

Home Connection 14 ★ Worksheet

Coins & Quick Sketches

Here is an array of quarters.

1 What is the total amount of money in this array? Use numbers, words, and/or labeled sketches to explain your answer.

2 Use the array to help solve these multiplication problems.

a 4 × 25 = _____   **d** 10 × 25 = _____

b 6 × 25 = _____   **e** 12 × 25 = _____

c 8 × 25 = _____   **f** 14 × 25 = _____

3 Rosie says she can solve 24 × 25 using the information above. Do you agree with her? Why or why not?

Home Connections

Home Connection 14 Worksheet (cont.)

4 Use what you know about adding and multiplying money to help solve the multiplication problems below.

example
$$\begin{array}{r} 25 \\ \times\ 36 \\ \hline 900 \end{array}$$
I know there are four 25's in 100 (four quarters in a dollar). 36 is equal to 9 groups of 4. So, 36 × 25 is like 9 × 100.

a $\begin{array}{r} 25 \\ \times\ 24 \\ \hline \end{array}$ **b** $\begin{array}{r} 25 \\ \times\ 32 \\ \hline \end{array}$ **c** $\begin{array}{r} 25 \\ \times\ 40 \\ \hline \end{array}$ **d** $\begin{array}{r} 25 \\ \times\ 34 \\ \hline \end{array}$

e $\begin{array}{r} 50 \\ \times\ 2 \\ \hline \end{array}$ **f** $\begin{array}{r} 50 \\ \times\ 16 \\ \hline \end{array}$ **g** $\begin{array}{r} 50 \\ \times\ 24 \\ \hline \end{array}$ **h** $\begin{array}{r} 50 \\ \times\ 32 \\ \hline \end{array}$

i $\begin{array}{r} 50 \\ \times\ 33 \\ \hline \end{array}$ **j** $\begin{array}{r} 50 \\ \times\ 17 \\ \hline \end{array}$ **k** $\begin{array}{r} 75 \\ \times\ 2 \\ \hline \end{array}$ **l** $\begin{array}{r} 75 \\ \times\ 16 \\ \hline \end{array}$

Students in an intermediate classroom have been working on multiplication arrays and money relationships (quarters/dollar) to estimate and solve for multi-digit multiplication problems. The assigned homework directly relates to the classroom work and provides examples for support.

These factors are the joint responsibility of the principals, math coaches, staff, and families at the school. Each school site must identify and prioritize the factors based on their community needs and take action on the work that will make a positive difference. Math coaches can help ensure their time and other resources are allocated toward that work. Many schools are working very hard, but not necessarily on interventions that increase student achievement. The staff at these sites may be discouraged at their lack of progress. Focusing on the key factors, on the "right" kind of work, generates forward progress.

Personal and Professional Qualifications of Math Content Coaches

Successful coaches are people who are compassionate, have a sense of humor, have high expectations of themselves and others, are able to actively listen, and are trustworthy, and thereby make a positive contribution to the school. These personal qualifications are certainly part of a coaching relationship.

The professional qualifications are equally important. Math coaches know the current research on educational leadership, including professional development for adults, student achievement, and mathematics instruction and assessment practices. They have demonstrated success using a standards-based curriculum and instructional materials effectively and are interested in learning more and making a difference. Math coaches have an understanding of K-8 mathematics and knowledge of its connections to higher levels of mathematics They are aware of the developmental landmarks in mastering early numeracy concepts, computational fluency with whole numbers, place value concepts, and fractions, to name a few. Math coaches use the NCTM Grade Level Focal Points and NCTM Content and Process Standards to focus their professional development work. They employ best practice for adult staff development.

For example, a math coach may choose to use the Grade Three Focal Points (Appendix 2) to investigate fractional relationships with a third grade team of teachers.

Number and Operations: Developing an understanding of fractions and fraction equivalence.

Students develop an understanding of the meanings and uses of fractions to represent parts of a whole, parts of a set, or points or distances on a number line. They understand that the size of a fractional part is relative to the size of the whole, and they use fractions to represent numbers that are equal to, less than, or greater than 1. They solve problems that involve comparing and ordering fractions by using models, benchmark fractions, or common numerators or denominators. They understand and use models, including the number line, to identify equivalent fractions.

NCTM Focal Points, Grade 3

During the course of multiple investigations with fractions as area models (geoboards and pattern blocks), linear models (rulers and number lines), and fractions as parts-of-a-set models (colored tile and egg cartons), coaches and teachers uncover the explicit mathematics and accompanying vocabulary that is addressed in the third grade Focal Point as learners themselves. In some cases, other master teachers may also facilitate a lesson to the participants using best practice teaching strategies. These explorations are used as an opportunity to identify misconceptions the students (and some teachers) have about fractions and their equivalence. For instance, thinking that ¼ is greater than ½ because the 4 (the denominator) is greater than the 2, or that ¾ is smaller than ⅜ because 3 and 6 are greater than 2 and 4.

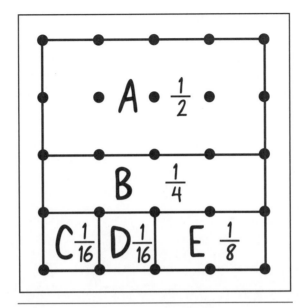

In this lesson, third graders determine the value of each area using a variety of proofs. They use fraction notation to represent parts of a whole.

Math coaches have a command of the district and state standards and benchmarks; therefore, they are able to facilitate an examination of assessment items used to determine proficiency with the third grade teacher team, as they align with this grade level Focal Point. They are also able to extend fraction concepts through the grades and help teachers differentiate their instruction for those students ready for additional challenges. Coaches know that adult learners are pragmatic. They want information they can take back to their classrooms and use with their students. Therefore, coaches continue to make connections to the classroom, planning for the fraction unit along with the teachers, focusing on instructional decisions, and shoring up the content knowledge of the participants through inquiry surrounding their grade level Focal Points.

Final Thoughts

Math coaching requires both these personal and professional qualities. Successful candidates must have an understanding and acceptance of diverse teaching styles and effective approaches to communicate with adults in addition to the mathematics knowledge to provide ongoing professional development. Math coaches influence the school culture, build and maintain a team, mentor other teachers as math teachers and leaders, and improve student achievement. They are positioned at the fulcrum for change—where the challenges are. Successful math coaches work to solve the challenges in an efficient and effective way. Because they have no formal power, they must find other ways to motivate, mobilize, and lead teachers. They rely on intrinsic leadership abilities, knowledge of mathematics and group dynamics, influence, trust, and respect.

2
Preparing Yourself

*"Courage means heart. It comes from the heart and must return
to it for guidance and renewal. It is not to be found in one great,
heroic act, but in day-to-day actions that come from the heart, and
from our willingness to take the path of heart."*
The Seven Acts of Courage, Robert Staub II

Coaching others requires courage, preparation, and thoughtful consideration of the challenges ahead. Therefore, math coaches work to be aware of what teachers do in the classroom, what the school community wants to do, and what they can do to make a difference. This requires an initial investment in preparation, and a day-to-day willingness to return to the heart for guidance and renewal. This chapter will examine effective leadership characteristics, help coaches identify specific goals for the school year, and suggest ways for scheduling time and communicating with others.

Effective Leadership

Many historians, politicians, and educators have studied leadership. Essentially, math coaches are content and pedagogy teacher leaders. However, the transition from classroom teacher to coach, as a teacher leader, is a difficult one for some educators. Shedding an old skin and growing into a new one requires a reflection on what makes an effective leader. The following characteristics were observed by Michael Fullan.

Characteristics of an Effective Leader

♦ Principled—stand and fight for what they believe is right and good for others

♦ Honest and ethical—trustworthy, maintain confidentiality

♦ Organized—have some kind of system in place, stay focused and on track

- ♦ Perceptive—sense what people need and WHEN they need it

- ♦ Empathetic and supportive—care and understand

- ♦ Accessible—have an open-door policy during the day, some lunches, free time to help others, are available

- ♦ Resourceful—have problem solving skills, know where to look for information

- ♦ Fair—hear all voices, even when they are different

- ♦ Accepting—accommodate and meet you where you are

- ♦ Vulnerable—admit mistakes, show a willingness to grow

- ♦ Forward thinking—know research and take calculated risks

- ♦ Global—see beyond the classroom and personal agenda

- ♦ Decisive and incisive—action-oriented and perceptive

- ♦ Intelligent—know the content and research and able to make connections

Classroom teachers, now math coaches, might want to identify the five characteristics of effective leaders that describe them best. *Why did they apply for the position? Why did the interview team choose them? What are their fears?* As they redefine their new role as a teacher leader, they might write down their reflections in a journal and keep them nearby. Some classroom teachers will initially grieve for the intimacy they had with a class of students and the camaraderie they enjoyed with other teachers; the safety of what they knew and did well. This grief is a natural part of the change process.

> *To use fear as the friend it is, we must retrain and reprogram ourselves. We must persistently and convincingly tell ourselves that the fear is here—with its gift of energy and heightened awareness—so we can do our best and learn the most in the new situation.*
>
> *Peter McWilliams*

Journaling provides an outlet for these thoughts and feelings and a record of a very personal passage from classroom teacher to math content coach. Developing and practicing effective leadership skills will take awareness, persistence, and energy.

Roles, Responsibilities, and Goals

Successful math coaches have clearly defined roles, responsibilities, and goals. These may come from the district administration, building principal, or leadership team. If they don't, math coaches create them before they begin the journey (see suggestions in Chapter One). Goleman (2005) identified a *coaching* leadership style as "a leader who develops people for the future." Great leaders move us, ignite our passion and inspire the best in us. Great leaders, according to Goleman, work through the emotions.

> *Their success depends on <u>how</u> they do it. Even if they get everything else just right, if leaders fail in this primal task of driving emotions in the right direction, nothing they do will work as well as it could or should.*
>
> *Daniel Goleman*

This means math coaches listen. They ask questions, focus on students and their work in the classroom, and guide teachers' pedagogy and content knowledge, while thinking about what people need emotionally. Content coaches balance and adjust the district agenda with staff emotional needs. They learn "how" and "when" to present information in a way that will work. Coaches set achievable goals for this school year and beyond.

Setting Goals for The Year

Coaches determine what the needs of the teachers, principal, and students are, based on a variety of data sources. The current reality defines what the math coaches will prioritize during the first school year. If something isn't working, the key players must be willing to say so. Perhaps the school needs to identify curriculum materials, gaps in implementation, and staff development needs. Ideally, building leadership, some classroom teachers, and the coaches come together to sketch an outline of a three- to five-year plan, that is both dynamic and measurable. In addition to identifying the needs, the leadership team identifies the assets and support structures. Math coaches may try writing goals from a school mission statement like the one below:

> *We will create a learning community in which all educators learn to support and learn from one another so that they can create high quality teaching and learning experiences for all students.*
>
> *School Mission Statement*

Although this statement reflects the mission of the school, it is difficult to assess because the declaration is not specific enough to be measurable. When writing goals, math coaches state a time frame, name a product, define the product, and state the purpose with a keen focus on student learning in mathematics. They ensure the administration allocates the resources for each of the goals, including time, money, and energy. Goals can be set for each quarter, or marking period during the year, and revised as needed. These goals should specify what will happen, who will make it happen, and how this goal fits into the overall plan.

Math content coaching goals for the beginning of the year may look like the following:

♦ Organize the math manipulatives available to teachers in the storeroom and purchase what is still needed to increase accessibility and engagement for all students

♦ Align the state and district math standards with the curriculum materials being used and identify any gaps

♦ Align the state and district math standards with the assessments being used and identify any gaps

♦ Model, observe, or team-teach a math lesson with each teacher in the building at least once this quarter

♦ Host a Family Math Night with a focus on standards-based teaching and learning.

Content coaching goals for the middle of the year may look like the following:

♦ Organize a book study for teachers around the Instructional Strategies (Marzano) that increase student achievement in mathematics

♦ Model, observe, or team-teach a math lesson with each teacher in the building at least once this quarter, with a focus on those Instructional Strategies

♦ Release teachers so they can observe each other during a math lesson and provide feedback on the use of the Instructional Strategies

♦ Host a Family Math Night with a focus on how families can help at home

Content coaching goals for the end of the year may look like the following:

♦ Select a professional article to share during staff meetings about differentiating instruction using content, process, and products and flexible groups based on interest, readiness, and learning profiles (Tomlinson, 1999)

♦ Present an adult level math lesson to the staff, focusing on the NCTM Process Standards-Problem Solving, Reasoning and Proof, Communication, Connections, and Representation (Appendix 1), and discuss how we can differentiate the instruction for a variety of learners

♦ Examine math assessment data with grade level groups, with an eye on making instructional decisions

♦ Observe or team-teach a math lesson with each teacher in the building at least once this quarter, with a focus on the use of differentiated instruction strategies

Using the instructional strategies of Identifying Similarities and Differences, coaches model the use of a Graphic Organizer to compare two primary students.

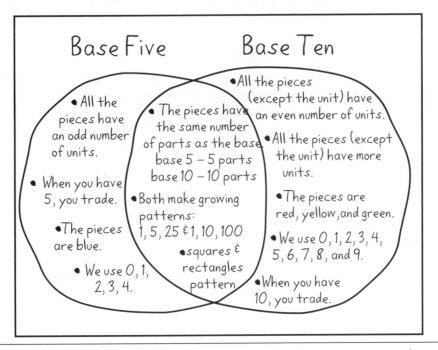

Intermediate students use a Venn diagram independently to compare and contrast base five to base ten structures in their math journal.

In each case, the goals are a positive statement of action, focused on student math achievement, and correlated back to the job description. They are specific and measurable and become the action plan for the coach. The record of each event (date/time and place), with accompanying student work or teacher/parent feedback becomes the evidence coaches need to reflect on their practices. This documentation also communicates the value of content coaching to the administration. With permission, math coaches can also use photographs, videotapes, and audio recordings of lessons and conferences. Below, a math coach reflects on her first year and the value of keeping records:

> *After completing my first year as a math coach I recognized the need to document what I had accomplished, not just for the administration, but for my own personal and professional growth. Sometimes, when I am just going and doing, it doesn't seem like I accomplish much. But when I stand back, I can see the whole picture and the progress we are making as a school community. Documentation keeps me focused on doing the right work and not just working hard.*
> *Math coach reflection*

Scheduling Time

Classroom teaching requires planning time, and math coaching does as well. Coaches need time to plan thoughtful lessons and meetings and reflect on their practice. This block of time can certainly be decided after the classroom teachers' math teaching schedules are blocked, but should not be ignored. Coaches need to stay motivated and energized by taking time during the day to catch their breath and think about the time spent with students, teachers, and the administration. Taking down some notes about next steps, perhaps in a journal or personal computer, can guide coaches in making thoughtful decisions, and not just reacting to the next crisis.

Coaches need time to pre-conference, teach, or observe a math lesson and post-conference with their teachers. A block of one and a half to two hours should be scheduled for each interaction. These blocks don't have to be continuous. For example, a coach might meet with a teacher in the morning, work in the classroom after recess, and post-conference during a specialist time in the afternoon. Ideally, these all happen in the same day, but if that's not possible, the post-conference can happen the very next day. Coaches keep these conferences informational and not evaluative.

Coaches need to meet with other coaches on a regular basis. Once or twice a month is recommended. Collaboration and reflection on coaching practice for the purpose of improving student learning must be the focus. It is important to vent about the responsibilities and frustrations, but equally important to move on. At these meetings, coaches are encouraged to contribute book reviews, articles and journals and share what's working with their colleagues. Coaches spend time reflecting on the needs of the adult learner, their motivation for learning, preferred learning methods, and staff development delivery because they recognize that adults want their learning to be problem-oriented and self-directed (Zemke 1984, Fogarty 2007). Ideally, the content coaches are meeting about the same content— mathematics. If that is not possible, meeting with literacy coaches or science coaches about the role of coaching is an option.

Scheduling time for planning, meeting with classroom teachers and other math coaches, and additional duties can be complicated. Early in the year, coaches may also be helping teachers who need support with managing manipulatives, centers, anecdotal records, lesson plans, parent contacts, and reporting results on mathematics assessments to students, parents, and the school or district. Being clear about the roles and responsibilities of math coaching, while being available and flexible, is one way to establish a positive relationship with colleagues.

Communication Strategies

Written communication needs to be friendly and professional, clear and concise. Math coaches may use a passive voice— where the emphasis is on the action or what people need to do. For example, *"The sixth grade team will be meeting at 12:00 Thursday in the library to discuss the district assessment data. Please bring your state standards and district benchmarks, class list, and an appetite. Pizza will be served."* This conversational style works well with e-mails or notes in mailboxes. Math coaches ensure their communications are brief, and proofread for errors and tone. *What attitude is present?* When teachers send an e-mail, coaches reply to it within 24 hours when possible, even if to say they are working on something else and will get back to them later in the week. When using e-mail, coaches are very careful not to make any private information public. E-mails are often forwarded to others and therefore cannot be secured.

Some coaches are selected prior to the following school year and are already known by most of the staff. In other circumstances, coaches are hired over the summer and new to the staff. These coaches may want to introduce themselves to the teachers over the summer through a welcome back letter.

During the school year, coaches can keep everyone up to date on the upcoming events and celebrations in quarterly newsletters. This written documentation is important as a body of evidence to garner teacher and administrative support and communicate the next term's goals.

Oral communication generally requires coaches to react to a situation immediately. For example, *"Why do we have to do this? This is a waste of time."* With some practice, coaches begin to respond to a question, with a question. *"How do you want to handle this?" "What if...?"* This gives coaches a bit of time to gauge the situation and pull it into a more positive realm. It's also acceptable to say, *"I'm not sure, but I'll find out."* Coaches try not to react to an emotionally charged situation when they are emotionally charged. They will also want to admit when they don't know, rather than make something up that is not exactly correct and have to retract the information later. Accomplished math coaches respond after they have thought about the concern, collected additional background information from building leadership and other stakeholders, and then made an informed decision or comment.

Dealing with Conflict

Difficulties are meant to rouse, not discourage. The human spirit is to grow strong by conflict.

William Ellery Channing

Many coaches avoid conflict because they don't like it. However, it's through the resolution of conflict that teachers grow in their content or pedagogy knowledge and understanding. The disequilibrium is a sure sign of growth and learning! When dealing with conflict, effective coaches follow a few general rules.

Dealing with Conflict

♦ Don't procrastinate, even if it's difficult

♦ Set up a time to meet with a colleague, and don't delay the discussion. The problem is not going to go away.

♦ Do make it nonconfrontational. Set up the situation so that the people involved can sit or stand side by side, and not behind a desk

At the meeting, coaches begin by saying something positive about the classroom or the curriculum topic. They may invite the teacher to share her/his thoughts. Then the coaches say what needs to be said, in a plain and succinct manner. If the teacher is using a mathematical term incorrectly or if the students are not engaged and making sense of the mathematics, the coaches must say so. The coaches may inquire about how the teacher sees the situation, and then brainstorm some possible solutions together. Finally, the coaches end the interaction with a positive comment about at least one thing that will change in the next lesson as a result of this conversation. For example, "I am looking forward to seeing you use word wall cards with definitions on the back, to clarify vocabulary terms" or "I am looking forward to seeing your students show and explain their thinking about decimals and percents next week". Coaches communicate enthusiasm for this change verbally and in nonverbal ways through the use of encouraging facial expressions and body language. They document the discussion that transpired and the result of the change initiative as a record of their coaching work.

Final Thoughts

When coaches begin to define the current reality in their school and set goals, they need to be thoughtful about what will happen and in what order. They cannot impact everyone, with everything all at once! The challenge for math coaches is to create and sustain the momentum for quality mathematics instruction, even as other building initiatives demand time and attention. Coaches focus on the math plan while making connections to other schoolwide initiatives. Their readiness to take the path of the heart in day-to-day interactions will guide and renew their journey.

3

Collaborating with Administrators

In *Good to Great* (2001), Jim Collins concluded that effective leaders "get the right people on the bus and the wrong people off the bus, and then work to get the right people in the right seats on the bus." He summarized that, "People are not your most important asset. The right people are." Schools that make a leap from good to great have leaders that work toward this goal. Administrators, principals, and content coaches are all part of this leadership team. When they work collaboratively, they transform schools from good to great. They tackle their current reality and maintain unwavering faith that good will prevail, regardless of the difficulties. Great coaches are the *right* people on the *right* bus, in the *right* seat, and they need to know who is driving the bus. This chapter explores the development of collaborative relationships among the leaders, especially the building principals.

Meeting with the Principal

Some coaches are hired from the teaching ranks in the building or district they serve and therefore already know the principal and staff. Their challenge is to form a new relationship with these same people based on their content coaching responsibilities. This association is forged from a fresh vantage point. No longer are these teachers responsible for a classroom of students. They are, at least in part, responsible for classrooms of teachers. Initially, in-building hires have a familiarity with the people they serve. This makes the job easier and more challenging. For example, if the principal has known the coach for 12 years and mentored him early in his teaching career, it may be difficult to shift managerial responsibilities and leadership duties to someone who is remembered as a "rookie teacher." In this case, the coach needs to demonstrate his professional growth by stepping up to the plate and managing a project successfully. This dispels preconceived notions and helps define a new relationship.

When coaches are hired outside the school community, they work to initiate a relationship from the ground up. These coaches have much to learn about the personalities in the school, the professional culture, and curriculum and instruction goals. They start fresh, with no connections, and make a valiant effort to get to know the administration team and teachers, before pursuing a course of action.

Regardless of the coaches' entry point, an initial meeting with the principal is in order. During this first meeting, content coaches find out about the principals' priorities for the school year and identify both formal and informal goals. In most cases, there are objectives that are written down and others that are implied. These are equally important to know and understand. As instructional leaders, principals determine the schedules, allocate resources, and evaluate teacher performance based on their own philosophy of teaching and learning. Principals, along with curriculum specialists, generally determine what professional staff development will be offered at the school. These responsibilities, along with the principal's personality, organization, and management style will directly impact the coaches' work. It is therefore essential that coaches and principals get acquainted on a professional level. Below are a few recommended questions coaches may ask their principals.

Questions to Ask Principals

1. What are your priorities for the school year? How can I help make these happen?

2. Are there other goals that are important for me to know about? How can I help you meet these goals?

3. What schedule changes have you made to accommodate these goals? Are there any additional changes you want to make?

4. What resources have you allocated for these priorities? What resources are still available?

5. What do you believe about teaching mathematics and student learning? What do you want to see happening during a math lesson in your building?

6. What professional development has been planned for this school year? How can I help support this plan?

7. Where do you think I should start?

Coaches may want to choose two to four questions to ask at the initial meeting, and a few to ask at another time, because this get-together should not feel like an interrogation! As they actively listen they'll get to know their principal's personality, student and family populations, and community culture. To learn more about others, coaches need to be talking less and listening more. After this first meeting coaches determine: *What are the math education priorities in this school? Does the principal embrace a standards-based philosophy of teaching and learning mathematics? Is the principal open to learning more about the research behind these initiatives? Does the principal expect the staff to align their teaching practices with this view? Where are staff members along a continuum in relationship to these expectations? How are the students doing on district and state assessments?* The answers to these questions will determine a plan of action.

Personalities and Possibilities

Coaches will be working with a variety of personalities, including the principal's. The principal's professional preparation in leadership and mathematics coupled with their persona will impact the content coaches' work. Although each administrator and school site is unique, coaches may want to consider the following possibilities:

Principal as math leader: When instructional leaders embrace standards-based mathematics teaching practices, the school priorities and resources are generally focused on increasing teacher pedagogy, math content knowledge, and student achievement. These principals are frequently in the classrooms observing and monitoring math instruction and attend math professional development right along with their staff. The leadership and most teachers at this site probably welcome content coaching as one more way to improve student learning and their teaching practices. Upon hearing the good news, content coaches at this site feel like they have hit the jackpot! The challenges that lie ahead are about making a good school great. Coaches at this site require more content and pedagogy knowledge because these people demand it. Chances are these schools still have some pockets of resistance or fragile links. Fine-tuning who will need what, how and when, helps forge a shared venture between the principals and content coaches. These principals want to be informed of the coaches' work and be a part of the decision-making process. In the spirit of collaboration, they'll continuously engage in professional conversations about the heart of math teaching and learning throughout the year.

Principal as a cooperative partner: Some principals are not aware of the best practices that depict "teaching to the mathematics standards." Perhaps they have been preoccupied in a literacy adoption, and because of that focus have let math instruction simmer on a back burner. They remember the way math was taught when they were in school. When these principals are in the classrooms during math instruction, they see a range of teaching practices that they aren't able to categorize as good or bad. They just don't know the difference. Coaches at these sites can begin by creating a vision of what quality mathematics instruction looks like along with the principals via articles, Web sites, curriculum resources, and ongoing professional development. Teaching practices, process and content standards, and methods of evaluation have changed as indicated by the Recommendations for Teaching Mathematics (Appendix 3). In order to increase our attention to best practice, teachers must decrease their attention to some of the traditional ways of teaching mathematics.

The use of manipulative materials, visual models (sketches), and cooperative partner work encourages the discussion of mathematics and prepares students for writing about their conjectures. Students build growing patterns with colored tile and explain the "rule" to their classmates. Then a chart is created to show the numeric sequence.

Arrangement Number	Total Tile
1	1
2	3
3	5
4	7
5	9
6	11
7	13
8	
9	
10	

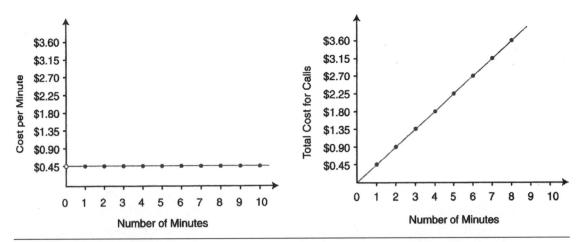

In middle school classrooms, students investigate several cell phone plans and represent quantitative relationships to determine the impact of a flat rate or cost per minute promotion. The real world problem-solving setting engages students in applying mathematics.

Math coaches model these recommendations during lessons in classrooms and during staff meetings and invite the principals to observe and reflect on these practices. They continue to engage the principals in discussions about additional resources and coaching priorities throughout the year. Great coaches don't settle for cooperative associations; they seek to create collaborative relationships with their principals throughout the year.

Principal as a disinterested party: There are also principals that are satisfied with the status quo and may even avoid a cooperative relationship with content coaches. They may see content coaching as "the latest fad or one more thing to do." Mathematics student achievement and instruction is generally a lower priority at these sites. In these situations, coaches may want to select several classrooms where the teachers are willing to implement standards-based instruction and develop collegial relationships with them. These classrooms are likely to show an increase in student engagement and achievement, and as a result, slowly get positive attention from the other teachers and principals. Mathematics instruction and content coaching may become a higher priority the following school year. In this situation especially, coaches must carefully document what they are doing in meetings, seeing in classrooms, and how they are impacting teachers and students. This data is used to communicate with the principals when they are ready to make inquiries and work toward a cooperative relationship. These principals commonly allow coaches more freedom to decide what they do, and how and when they do it, and most coaches enjoy this autonomy. But that freedom comes with additional responsibility for student achievement and less administrative support if something goes wrong.

These three scenarios only begin to describe the possible combinations of interactions between coaches and principals. Each school site presents its own challenges and opportunities. This relationship can be enhanced by looking at assessment data, observing teachers, and conferencing about strategies that increase student achievement.

Dialogue and Documentation

For content coaching to make a significant impact, there needs to be ongoing conversation between coaches and principals. This dialogue provides the opportunity for collaboration, learning, and planning. In the fall, principals and coaches assess the reality of the school climate and set goals for their collaborative work. These goals include data inquiry, setting up schedules, identifying preferred communication methods, and professional agreements about supporting each other as leaders. They may also map out the staff development that will be offered during the year and who will be responsible for organizing and presenting the information to the staff members.

During the school year, the principals observe the coaches teaching and provide feedback at a post-conference. The coaches continue to document their work with teachers and students, specifically as they relate back to the goal-setting session in the fall. Together, they team to resolve problems as they occur. Ideally, they meet at least twice a month to examine progress made towards the goals set at the beginning of the year. Questions like *"How are we meeting the needs of the staff? How are the one-to-one, small, and whole group professional development models supporting effective teaching and learning? What are our next steps?"* are discussed and resolved during these bimonthly meetings.

In the spring, coaches and principals assess their impact on schoolwide achievement and completion of the goals set in the fall. They articulate what was done, how well it was done, and what to do next. They examine achievement data from several sources including anecdotal notes from in-service and coaching sessions. Along with the entire staff, they celebrate their successes.

Gathering Data to Drive Instructional Practices

With the passage of federal and state accountability measures, student and school performance has become imperative. Schools collect and manage a wide array of informal and formal data. This data comes from families, previous teachers, classroom assessments and observations, district and state tests, and national and international studies. Along with principals, one important role content coaches might assume is an analysis of this data. Inundated by data sources, they must sift out what is useful, for what purpose. Having the right data available at the right time makes decision making purposeful. *Who will use this data? What does it tell us? What is missing?* Content coaches proactively plan for this responsibility by asking to see school data and preparing for these conversations.

Data Homework for Coaches

♦ Identify existing sources of data (multiple sources) used by the school site

♦ Identify direct and indirect sources of existing data used for making decisions

♦ Examine existing data charts from the school site and the district, state, and national levels, especially those shared with families and community members. Write down questions about what you see and don't see in the charts

♦ Identify ways to disaggregate existing math data by gender, attendance, socio-economic status, and special education subgroups

♦ Look for patterns, interpret and analyze trends in existing data; *What do you notice? What do you wonder about?*

♦ Identify data connected to any math interventions being used at the site. Interpret the results. *Are these interventions making a difference in student achievement? Are other data sources needed?*

Discussions about data will most likely relate to the student performance standards set by the state and run parallel with the school mission and goals. Building leadership may ask questions about the characteristics of students making the most dramatic gains or losses on district, state, and national tests. *How does attendance impact students academically? How well do second language and special education students achieve? Are they able to show growth?* These questions and others launch a collaborative investigation. To draw conclusions about student achievement, multiple sources of data need to be examined from a variety of settings and show student progress over time. Standardized state tests given once a year are balanced with quarterly, monthly, and weekly data to measure how students are doing.

> *Receiving test data in July is like driving a school bus looking out the rear-view mirror. I can see where my students have been but I cannot see where we are going.*
>
> Dr. Gregory Decker, Raleigh, NC

Analyzing data alongside administrators, coaches begin to see a bigger picture of student achievement — state test scores, district assessment grades, school report cards, rubric scores of student performance, and records from computer-managed software. This body of evidence makes it possible to determine which skills students have mastered or still need help with, identify groups of students who need special interventions, draw conclusions about mathematics curriculum standards/areas that are particularly strong or weak, and shape instructional programs based on this achievement data.

When examining data, coaches may also notice most teachers embrace their own classroom observations and assessment practices as compelling. They believe what they see and hear in their classroom on a daily basis. They are sometimes suspect of national or state-level exams because their students are working in a format that is not familiar to them or tested on items not taught at their grade level. Administrators often embrace national and state achievement test data because they are "standardized" and have gone through extensive assessment review. They are sometimes suspect of teacher observation and assessment practices because they are not consistent. *Which assessments do teachers value for making instructional decisions? Which assessments do principals value for making decisions? Why?* These professional conversations related to data analysis may take place during the year, as coaches begin to inquire how to best measure student achievement in mathematics.

Principals want to know if teachers are preparing students for the content and format of the test and if teachers and students are taking the tests seriously. If they are not, they will want to know what the coaches can do about it. Both parties may want to attend professional workshops on assessment and data analysis together to establish a common language and structure for data-driven decision making. Math coaches might create pacing guides for instruction to encourage teachers to teach the grade level content prior to the assessment window. They may identify or create common grade level formative assessments and scoring guides. Coaches may also collect and organize this data into spreadsheets and make time for looking at assessment results with grade level teams. These additional assessment responsibilities ultimately contribute to increasing student achievement because they articulate what will be taught, when, and how it will be assessed, in an organized way. Data drives instruction. Teachers are more likely to make a change in their instructional practice and principals are more likely to support math content coaching when the data shows an increase in student math achievement. (See example of Post-Assessment Checklist on opposite page.)

Observing Coaches in Action

At many school sites, coaches model classroom lessons on a regular basis. One way to establish a collaborative relationship with principals is for coaches to invite the principal to observe a pre-conference, observe them modeling a lesson, and stay for the post-conference session. If all three invitations seem a bit overwhelming, coaches may begin with an invitation to observe a classroom lesson. Principals are frequently responsible for formally evaluating the coaches in their building, and therefore this observation fulfills their supervisory role. Before the principals come in, coaches brief them on the mathematics lesson. *Is this an introduction to a topic, an additional opportunity for practice, or a mastery competency for this grade level?* Coaches also specify which instructional strategies they will use during the lesson and why they chose to focus on those strategies. If the principal does not use a standard observation form, coaches may suggest on what they would like to be observed, for example, using higher- level questions to create student discourse about perimeter.

Unit Five Post-Assessment Class Checklist

Student name	Points	Sara	Kyle	Brian	Dan	Ginny	Izabella	Adam	Alex	Pablo	Adri	Rachell	Kaitlynn	Kendra	Lucus
1a Lists data points in order from least to greatest	1*	1	1	1	1	1	1	1	1	1	1	1	1	1	1
1b Finds the range, mode, and median of a data set	3	3	2	3	3	3	3	3	3	3	3	3	3	3	2
1c Creates a well-labeled bar graph	5	5	5	5	5	4	5	4	5	4	5	5	4	5	5
1d & 1e Generates questions that can and can't be answered on the basis of the graph	2	1	2	1	2	1	2	2	2	2	2	2	2	2	1
2a & 2b Finds the mean of a data set and shows work	2	2	2	2	2	2	2	2	2	2	2	2	2	2	2
3a Interprets a circle graph	1	1	0	1	1	1	1	0	1	1	1	1	1	1	1
3b Reports the correct answer	1	1	0	1	1	1	1	0	1	1	1	1	1	1	0
4 Identifies the probability of 5 events using terms such as *certain* and *unlikely*	5	5	3	5	5	5	5	0	5	5	5	5	5	3	1
5a Identifies the probability of pulling a green tile out of a bag that has 5 greens, 2 reds, and 3 blues	1	1	1	1	1	0	0	0	1	1	0	1	1	0	0
5b Expresses probability as a fraction	1	1	1	1	1	1	1	0	1	1	1	1	1	1	1
5c Explains answer by referring to the number of tile in the bag	1	1	1	1	1	1	1	1	1	1	1	1	1	1	1
6a Identifies event as impossible	1	1	1	1	1	1	1	1	1	1	1	1	1	1	1
6b Identifies a number between 0 and 6	1	1	1	1	1	1	1	1	1	1	1	1	1	1	1
6c Explains answer to part b by referring to initial conditions and/or probability rather than luck or preference	1	1	1	1	1	1	1	1	1	1	1	1	1	1	1
7a Determines the probability of getting 3 heads when flipping 3 coins at the same time	1	0	0	0	1	0	0	0	0	0	0	1	1	0	1
7b Reports the correct answer	1	0	0	0	1	0	0	0	0	0	0	1	1	0	0
8a & 8b Selects the best spinner and explains choice in terms of the probabilities involved	2	1	1	2	2	1	1	1	1	1	2	2	2	2	1
**Total score / Level of proficiency	30	25	21	26	29	23	24	17	26	25	26	29	28	24	19
		P	B	A	A	P	P	N	A	P	A	A	A	P	P

* The total possible number of points for each problem is shown. ** **A** Advanced (working above grade level) 27–30 points (88–100% correct) **P** Proficient (working at grade level) 23–26 points (74–87% correct) **B** Basic (working toward grade level) 19–22 points (61–73% correct) **N** Novice (working below grade level) 18 points or fewer (60% or less correct)

In the checklist above, student proficiency with the Data and Probability Post-Assessment is tracked by specific test item. Careful analysis reveals that most students are proficient with finding range, mean, median and mode, creating and interpreting graphs (bar and circle), and analyzing the probability of a tile drawing. Few students were able to accurately determine the probability of getting three heads when flipping three coins at the same time. Given this assessment data, classroom teachers might revisit the coin-flipping investigation and subsequent outcomes again to clarify the confusion.

Pattern Block Perimeters   Lesson 18

⊕ Follow-up Student Activity 18.1

NAME _____ DATE _____

1 In your own words, explain the meaning of perimeter.

2 The first 3 trains in a sequence are shown below. Questions a)-d) refer to this sequence:

a) If this pattern continues, describe what the 10th train looks like and how you decided this.

b) What is the perimeter of the 10th train, assuming the length of one edge of a small parallelogram is 1 linear unit? Explain how you "see" the perimeter of train 10:

c) Explain how you "see" the perimeter of the 100th train.

d) Which train has a perimeter of 46? Explain how you determined this.

(Continued on back.)

Lesson 18   Pattern Block Perimeters

Follow-up Student Activity (cont.)

3 The first 4 trains in a sequence are shown below.

Find the perimeter of the 20th train and explain 2 different ways to "see" this value.

4 The first 4 trains in a sequence are shown below.

Describe a method (other than building the train and counting) that works for finding the perimeter of every train in this sequence.

5 Describe what is easiest and what is hardest now for you about working with visual patterns.

6 Describe one or more ideas about visual patterns that used to be hard for you, but are easier now. Tell what you think helped you to understand better.

7 Use your pattern blocks to form the pattern in problem 4 and show it to an adult. Find out their ideas about what the 100th train looks like and what its perimeter would be. How do their ideas compare to yours? Explain on another sheet of paper.

In this lesson students are given a collection of pattern blocks to solve for the perimeter, using non-linguistic representations and cooperative learning structures. Then, they complete the record sheet independently using the rhombus block. The questions encourage students to generate and test their hypotheses about perimeter and construct a formula for their thinking. Additional challenge is offered to interested students, using the hexagon block. Once students complete the assignment, the coaches facilitate the student discussion with questions like *"Will it always work that way? Can you convince us? What if the shape was an octagon?"*

The observation is then followed up by a post-conference with the principal. At this time, the principals ask questions and provide the coaches feedback on what they saw during the lesson. Coaches listen carefully for common ground to begin a professional dialogue related to implementing effective instructional practices, in this case higher level questioning. Questions like *"What's the answer to problem 1, 3, 5, and 7?"* are being replaced with *"How did you solve that problem? Is there another way?"* Coaches might share how they are also working with teachers to uncover students' mathematical reasoning and proof and encouraging students to communicate a variety of problem solving strategies. The pattern blocks provide a visual model for their thinking—a proof. This conversation might provide the principals with new understanding on best practices and the coaches' role with teachers in the building. What the principals notice and share with the coaches offers great insights about their values and beliefs. Content coaches are generally borrowing other teachers' classrooms for their teaching segment. This may present additional challenges related to classroom management structures and protocols established by the actual teacher. When principals provide feedback, some of their comments may be related to these challenges and how the coaches deal with them on a regular basis.

After the initial modeling observation, coaches may want to schedule classroom walk- through observations during math class, alongside of the principals, looking for good things to comment on and ways of promoting a professional learning community in the school. Perhaps clear and precise math language is heard, or classrooms have student- generated workup uncovering big ideas in mathematics, or teachers and students are using a variety of math tools to illustrate their thinking. These shared experiences provide background knowledge to draw on in planning the coaches' next steps. Later in the school year, coaches invite the principals back to observe another pre-conference, modeling lesson, and post-conference session. With teacher permission, math coaches may also invite the principal to observe a pre-conference, teaching lesson, and post-conference session and provide feedback on the coaching relationship between colleagues. Once again, the ongoing dialogue is an opportunity to examine the coaches' role and the shifts in values and beliefs. It provides a road map for future goal setting and follow-up work.

Taking Part in The Practice

Ideally, every teacher in the school should be asked to participate with the math content coaches in some manner. Teachers can observe a model lesson, team-teach with the coaches, or be observed by the coaches each quarter or marking period. This arrangement allows everyone to take part in the practice, at their comfort level, and avoids identifying the few teachers that may need the most assistance, but are most reluctant. Math content coaches can begin to schedule these classroom visits right away rather than waiting for willing teachers to contact them for an invitation.

Principals may also identify candidates for additional coaching opportunities based on their professional performance or novice teaching status. If so, principals should participate in the observation process and document their own evidence of proficiency because a content coach is not in an evaluative or supervisory role. At the end of the year, the coaches' observations and comments cannot stand alone and classroom teachers are made aware of those limitations.

In the privacy of the principal's office, principals need to be able to talk to the math coaches candidly about teaching practices in their building. This information is professional, extremely confidential and must never be discussed with other staff members. Coaches and principals must be specific about what math pedagogy and content needs to be addressed with teachers and set aside blocks of time to accomplish these goals. Math coaches report back to the principals on a regular basis using objective data to monitor their progress. For example, "*Ms. S is implementing a cooperative learning strategy in her lessons. She is teaching session ten in unit three, which is 17 lessons behind her grade level team. We are meeting to discuss the pacing issue on Thursday.*"

Staying Out of The Middle Position

Teachers vent to coaches about things that frustrate them about students, parents, and the administration. Coaches must be careful not to share those stories from the teaching rank. Coaches listen and are understanding, but not the direct conduit to the principals. If teachers are disturbed by something, they must tell the principal themselves, "*I want to quit! I am going to retire before I do this program.*" These messages, repeated by the coaches, will put them in harm's way. In some cases, after a cooling-off period, teachers will retract what they said, and coaches who have repeated what was said will lose their credibility with the teachers and principals. Coaches care about individual teachers and, like the school secretary, are the eyes and ears of school morale. They must practice discretion about when and how to raise issues that need to be addressed by the principals. At that time, coaches must ask themselves several questions, and answer them honestly, before bringing a topic up to the principal.

Questions to Answer Before Talking to a Principal

♦ How important is this issue to my teachers or to me? Is it my issue or the teachers? If it's not mine, should the teachers be doing the talking?

♦ What is the root cause of this problem? Should I gather more information about the history of the situation before moving ahead? Where can I get objective information?

♦ How will this affect my relationship with the administrator? And other staff?

♦ What's my purpose? Am I doing this for the right reasons?

♦ Will my involvement complicate or help matters?

♦ If I choose *not* to pursue the issue, will I still be respected as a professional and a teacher leader?

In some situations, math coaches may see or hear something and want the principal to take action immediately. For example, a teacher is not teaching the current math curricula but continually replacing it with worksheets copied from the previous program materials — supplanting the instruction. As coaches, they may not have a complete understanding of all the circumstances or the ramifications of the personnel issues. The math coach's inquiry might be, *"What are we going to do about this?"* The principal may not respond instantly. Perhaps this teacher is already on an improvement plan, and the principal discussed this issue with the teacher specifically on another occasion and is in fact monitoring the situation. Principals are not always able to discuss legal, contractual issues and confidential information with the coaches. It may seem like the principals are not listening to the coaches, when in fact they are listening but are not in a position to act on what is shared at that time. Math coaches may have to leave the information with the principal and trust that it will be dealt with at the administration level.

Owning Up to Mistakes and Mishaps

Occasionally coaches will have a situation brewing with colleagues. They need to discuss it with their principals so they don't get blindsided and own up to their mistakes or mishaps. Coaches must be forthcoming and honestly communicate what occurred and why. Principals will only be able to support the coaches when they are informed. For example,

> *Mr. T and I had a heated debate about the use of calculators in his classroom. He was very frustrated that many of his grade four students were not able to subtract four digits accurately and wants to stop and teach that skill instead of moving forward in the unit on data and probability. I asked him to read this article from the NCTM journal on the use of calculators, and I'd like you to have a copy as well. If students are able to understand the problem-solving situation, but are making computation errors, it seems like an appropriate accommodation for this strand. I agree that his students need additional practice with multi-digit computation, but in this setting calculators would help all students be successful in this data unit. I think this calculator issue may come up again, and I want you to be aware of it.*
>
> *Math coach reflection*

Final Thoughts

Being an instructional leader, reliable and hard working, will go a long way toward developing a positive rapport with principals. When coaches demonstrate competencies with a variety of tasks and their willingness to be challenged, principals become even more interested in collaboration. This relationship is strengthened because the coaches focus on math learning and the principals focus on the school community as a whole. Managing change in such areas as scheduling, budgeting, and supervision becomes even more important when there is a larger change initiative on the horizon. Therefore, it is critical to the success of a coaching relationship to have administrative input and leadership. The principals, not the coaches, are ultimately responsible for developing and articulating the changes into practice. This includes providing and participating in essential professional development and ensuring that implementation of the standards-based math curricula is guaranteed and viable. Finally, the principal is responsible for assessing and communicating the results of these changes to the district and state administration. Although math content coaches help articulate what needs to happen, and continue to work toward the implementation of these changes, the buck stops with the principal.

> *I worked with a district that set out a wide net of professional development opportunities for their teachers including support for the National Board for Professional Teaching certification, a university-sponsored cohort for K-8 math/science leadership, graduate-level math classes on weekends, a book study focused on standards-based math teaching, and hiring math coaches for each building. With all these initiatives, teachers began to hear the same message about research and best practice in mathematics. This call to action gave teachers choices about how to get on board and sent the message loud and clear that change was necessary to improve student performance in the classroom and on the district and state assessments.*
>
> *Math coach reflection*

Effective leaders build coherence in the midst of change, collaboratively; they transform schools from good to great. They recognize that people are their most important asset and invest in the process of getting the *right* people on the *right* bus, in the *right* seat.

4

One-to-One Collaboration

*"We cannot teach people anything. . .
we can only help them discover it for themselves."*
Galileo Galilee

At the center of math content coaching is the development of a professional relationship with individual teachers. This connection happens in a personal context and therefore creates an optimum environment for discovery; self-realization, and self-improvement. Much like working with diverse student populations, coaches work at differentiating the content, process, and product for their staff (Tomlinson, 1999). *What can they help the teachers discover for themselves? How can they best present new ideas to these individuals?* Each person, each circumstance, is unique and therefore requires careful observation and planning. Chapter Four addresses several types of one-to-one coaching relationships, provides suggestions for pre-conference planning and post-teaching conference feedback sessions, and offers suggestions for using classroom observation guides included in the appendix.

Beginning With a Positive Relationship

Empathetic coaches with effective communication skills create trusting and respectful relationships with their peers. These coaches constantly analyze and monitor their own actions and reactions and seek to understand the actions and reactions of others. Teachers need to believe that something significant will come from their interaction with the coaches; perhaps they want to improve their instruction, grow in their mathematical content knowledge, or increase achievement for all of their students. Coaches start with what the teachers have self-selected as a goal for the relationship and stay "tuned in" to what the teachers value. They are on the lookout for that opening, that moment when a teacher says, *"I wonder if...? I wonder why...? Do you think...?"* It's at that moment they establish contact

37

for the coaching journey. Initially, it may be a casual conversation in the hallway, lunchroom, or after school in the parking lot. Later it becomes more formal with a classroom observation or a peek at their student work.

Math coaches establish norms for working together collaboratively in a one-to-one relationship. They may ask individual teachers to specify what they want the coach to do, and never do, in their classroom or in another professional setting. For example, the classroom teacher may invite the coach to step in during a lesson if they notice students are stuck or need a challenge. Teachers may also request that the coach never interrupt them in front of their students, but provide them feedback privately. Some teachers will have an open-door policy for coaches to stop by anytime, while others may prefer, at least initially, for the visits to be scheduled ahead of time. This preliminary conversation helps create healthy boundaries and avoids hurt feelings down the road. Early on, coaches communicate any limits to their relationship, perhaps in experiences, resources, and time. They also share their own personal enthusiasm for establishing a connection and building their own knowledge. There are at least three kinds of coaching arrangements that are worth careful examination: a resource, modeling, or collaborative relationship.

A Resource *Relationship*

In this relationship, math content coaches are a resource to teachers. The coaches provide materials and information, and might even work cooperatively with the teachers in planning standards-based lessons or units. Teachers are excited to get new books, games, and math manipulatives for teaching concepts from the coaches. (See opposite page for game example.)

Math coaches might give teachers assessment data to review and articles to read. Their conversations are focused on what the students seem to understand (or not understand) and leave the teachers free to interpret that information. The coaches are not in the classrooms during actual instruction, and therefore don't have an opportunity to comment on the teachers pedagogy or the mathematics content. Coaches may initiate reflective questions for the teachers to think about: *What are you currently doing that is working? What manipulatives and models have you used? Which students are having difficulty? What do you think is getting in the way?* This is often the first phase of a coaching relationship the coaches are meeting with teachers in the *hall*, and not in the classroom during instruction.

A Modeling *Relationship*

In this stage of the relationship, coaches model what standards-based mathematics instruction looks like, sounds like, and feels like in the classroom. The instruction actively engages students in cognitively demanding tasks with coaches as the primary instructor. The lessons exhibit best practices that are differentiated to meet the needs of a variety of learners. When coaches teach and teachers

Fill It First!

Each pair of students will need

★ 2 Fill It First! gameboards
★ 1 Fill It First! spinner
★ 2 sheets of paper pattern blocks

Instructions for Fill It First!

1 You and your partner will need to get a spinner and two sheets of paper pattern blocks to share. Each of you will need your own gameboard.

2 Take turns spinning the spinner. Each time you spin a shape, take a pattern block of the same shape and place it on your board..

3 The first person to fill the apple on his or her gameboard wins. The catch is, you have to go out exactly. If all the space you have left is a rhombus and you spin a trapezoid, you miss your turn and have to try for a rhombus or a triangle the next time around. Continue playing until one person fills his or her apple entirely.

Instructional Considerations for Fill It First!

Here are some things you might look for as you watch students play this game and listen to their conversations.

- Can children name the shapes?
- Do they recognize that some shapes fill the puzzle more quickly than others?
- Can they make the needed flips and rotations to make shapes fit into the triangular guidelines on the gameboards?
- Are they able to take turns and wait patiently as their partner finds his or her blocks and sets them on the gameboard?
- Are they able to place each shape they collect in places that will give them some flexibility toward the end of the game?

2R Fill it First! Gameboard

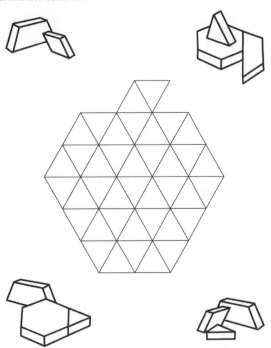

In the game **Fill it First**, from *Bridges in Mathematics,* kindergarteners practice naming geometric shapes (triangle, rhombus, and trapezoid), using spatial problem solving, taking turns, and sharing materials. The **Instructional Considerations** provide some suggestions for the classroom teachers, observation look-fors, that might foster further conversations between the teachers and coaches.

observe, the risk level is lower for the teacher. The coaches generally suggest follow-up lessons for the teachers to practice the strategies discussed during the modeling. The transition from observation-to-implementation generally takes several demonstrations over a period of time. Teachers need to "picture" and "practice" a new technique, supported by ongoing feedback, consciously working on the new skill until the approach becomes natural, and almost unconscious. At this time, math coaches are meeting regularly with teachers and listening to their response, gradually releasing the responsibility for the implementation and monitoring onto the teacher.

The demonstration lessons are carefully planned by the coaches. The teachers are usually assigned specific duties or areas to focus on. This is important because some teachers won't be able to resist jumping in to rescue struggling students or managing behaviors they deem inappropriate, and may miss the instructional value of the actual lessons. Some teachers may even jump on their computer to check e-mails during the teaching period!

After the lesson, the coaches and teachers reflect on the goals and experiences during a post-conference. For example, if increasing student math language was the identified goal, coaches and teachers would name the vocabulary words that would be likely to come up in the lesson. They would answer the following questions prior to the lesson: *What experiences and models would encourage all students to use these terms? What cooperative grouping structures would allow more students to talk more, to practice using these terms in the context of the lesson?* During the lesson, the teacher would collect data on the use of vocabulary terms by the coach and the students. *What did the coach do to scaffold the use of the clear and precise math vocabulary?* During the post-conference, the team would examine the impact models and grouping had on vocabulary development. *How did this practice impact student engagement and learning during the lesson? Were all students, including the English Language Learners participating? What other structures and resources would continue to support this vocabulary development?*

In the second phase of a coaching relationship, the coaches are in the classroom, working in the trenches with the teachers, in their personal and professional space. This requires sensitivity to the needs, wants, and values of the teachers and an ability to diagnose the next steps accurately. Math coaches that wind up the students and then leave the classroom may not be invited back. Carefully thinking about the teachers' comfort with noise, level of activity, and transitions at the end of the lesson, will help get the coaches a second invitation. Math coaches must have their radar tuned to the individual classroom teachers' responses.

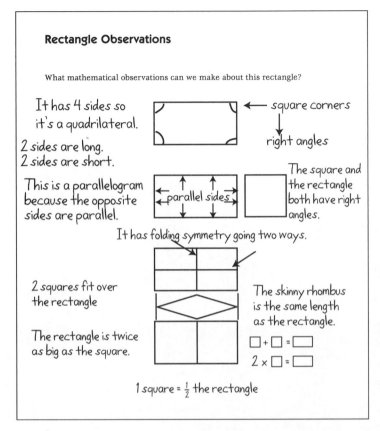

In this lesson, fourth grade students are given index cards and asked to share their mathematical observations about rectangles. The fact that the rectangle is white or made out of paper is not relevant to the properties of rectangles. The coach models how to journal about their thinking using precise math language and diagrams. Then students are asked to construct posters with peers about other polygon shapes and present their work to the class. While the coach is facilitating the student presentations, the teacher is able to take notes about the use of math vocabulary.

A Collaborative *Relationship*

A collaborative relationship develops between two people who share the same pedagogical beliefs about teaching and learning. In some cases this occurs quickly. Perhaps the math coaches and classroom teachers participated in the same undergraduate or graduate studies, took on curriculum and instructional work for the district, or went to national or regional math conferences together. These earlier experiences provide a base for launching a collaborative coaching relationship based on mutual trust and respect and similar values. In the absence of these previous contacts, math coaches and teachers develop this relationship over time. Each positive hallway contact, discussion at a staff meeting, shared classroom observation, or interaction during a professional development day provides an opportunity to cultivate this relationship.

A middle school teacher and coach want to focus on the communication standard, via problem-solving tasks. They agree that the content at this level is generally more complex, nevertheless some students continue to give just procedural descriptions of their process rather than analyzing their strategies. After reading about this topic in depth, they conclude that during this series of lessons they will

Bridges Student Book For use in Unit Three, Sessions 6 and 7.
NAME _____ DATE _____

Pattern Blocks & Angle Measure page 1 of 2

A straight angle measures 180 degrees.

1a Place 2 square pattern blocks together as shown below.

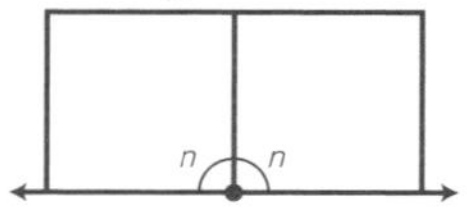

180 degrees

b Determine the value of each angle marked n. Show your work below, using sketches, numbers, and words.

c The value of each angle marked n is _____ degrees.

2a Place 3 triangle pattern blocks together as shown below.

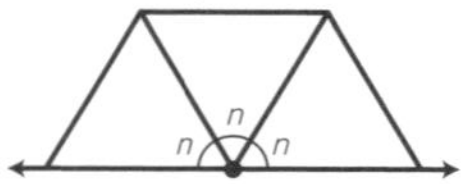

b Determine the value of each angle marked n. Show your work below, using sketches, numbers, and words.

c The value of each angle marked n is _____ degrees. (Continued on back.)

Bridges Student Book For use in Unit Three, Sessions 6 and 7.
NAME _____ DATE _____

Pattern Blocks & Angle Measure page 2 of 2

3a Place 6 white rhombus pattern blocks together as shown below.

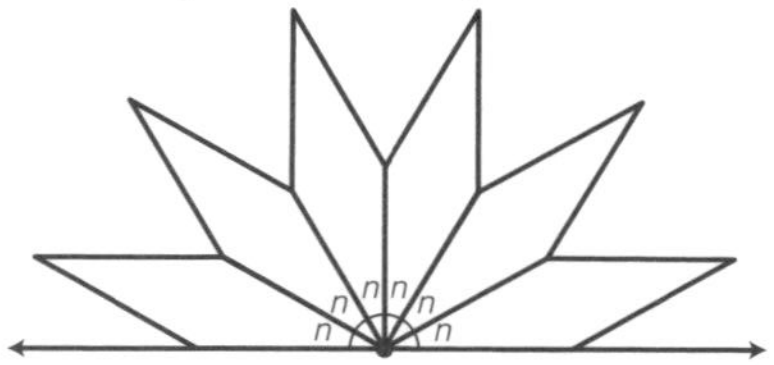

b Determine the value of each angle marked n. Show your work below, using sketches, numbers, and words.

c The value of each angle marked n is _____ degrees.

4a Determine and label the angle measures of the other three pattern blocks in the set.

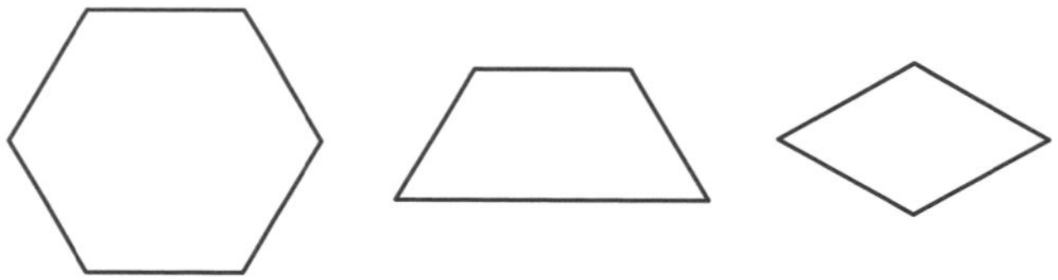

b Circle one of the shapes above and use words, numbers, and labeled sketches to explain how you figured out the measure of each angle in that shape.

In this session, students use pattern blocks to determine the sum of the interior angle measurements of polygons. They are encouraged to communicate their thinking using sketches, numbers, and words.

establish classroom norms that encourage communication, solicit more student-to-student questioning and encourage students to diagram their verbal explanations more carefully. In this task, students will not be allowed to use protractors (using tools and following procedures). (See example on opposite page.)

During this lesson, the classroom teacher monitors peer-to-peer discussions and facilitates whole-group sharing with the content coach observing and taking notes about the classroom norms, student questioning, and diagrams. After the lesson, they share their experiences, evidence, and conjectures.

- ♦ What did the teacher say or do to create a safe classroom community? He nestled in during the work time and actively observed his students working on the task. He also provided additional wait time and respectfully listened to each pair of students while they were sharing, modeling what he wanted the students to do during the presentations.

- ♦ What did the teacher do during the work session? He was careful not to jump in and show and tell the students what to do with the blocks and their diagrams. He asked questions (What do you notice about those two blocks? How can you show your thinking? What labels might help others understand your work?) that encouraged students to clarify their thinking.

- ♦ What happened during the presentations? Students were comparing their strategies to the others. "We used the triangle for our base piece instead of the rhombus." Other students noticed when the angle labels were not given and asked for more clarity. Students were making generalizations about the interior angles of other polygon shapes based on the total of these basic pattern block shapes.

In this stage, the coaches and teachers mutually support and promote each other, as they examine what is happening in the classroom, based on what the students and teachers are doing. They are practitioners engaged in deep, thoughtful, and insightful conversations regularly. The coaches and teachers challenge each other, to fine-tune their teaching craft by asking each other why they did this or that, or what if something else happened, then what? Communication about mathematical content and pedagogy is welcomed because they are co-learners and maybe even co-leaders in the building. The coaches and teachers have become collaborative, professional problem solvers. Their interdependence creates a dynamic learning environment for their students. This is the third phase of the coaching relationship the *heart* of teaching and learning.

Inquiry Continuum

Whether coaches are in the *hall, classroom,* or *heart* of the teachers they serve depends on two people: the coach and the teacher. The math coaches' personal and professional skills and their ability to engage teachers in the inquiry of teaching will impact the quality of these relationships. At any given time, math content coaches will be managing all of these relationships, with multiple groups of teachers. Knowing where the classroom teachers are in the continuum and responding accordingly is key to the success of a coaching program.

Does the classroom teacher avoid the coach by being "too busy," consistently overwhelmed, or absent on the scheduled coaching days? Does this teacher want the coach to help them by taking the struggling students out of the classroom for an intervention time? Is this teacher generally not prepared to discuss the assigned readings during staff meetings and professional development days? This reluctance must not be excused and ignored, but dealt with in a fair and systematic way. Math content coaches and principals expect everyone in the building to participate in the coaching initiative and hold everyone accountable on a regular basis.

Does the teacher want another activity to do with her students to make sure they "get it"? Is the teacher interested in having the coach "show and tell" a lesson? Is this teacher unsure of the developmental stages in this area of mathematics? For example, how do students construct an understanding of place value? Is the teacher willing to read current research, try out a few new ideas, and come prepared to the meetings? When teachers are approachable, math content coaches can make a significant impact on their practice especially when they see evidence of student engagement and achievement.

Does the teacher want to be observed, and get feedback on how to get better? Does the teacher bring in a pile of student work and want to examine what the students know and can do? Does the teacher find articles and Web sites on best practices, math content, and pedagogy to share with the coach?

Coaches are seeking the answers to these questions to determine who is ready, for what content and pedagogy, and when. They are enthusiastically working towards getting into the classrooms during instruction. When an appointment is set, coaches will want to make time for a pre-conference with the classroom teacher.

Planning a Pre-Conference

The function of the pre-conference is to frame the math lesson for the teachers and coaches. It is time to clarify the lesson and the mathematics that will be taught. It provides the coaches with insights to the teacher's depth of mathematical knowledge and can also provide time for discussion on effective instructional strategies (Appendix 4). The pre-conference is a good opportunity to brainstorm any possible challenges students may have and fundamental misconceptions that may be addressed up front in the lesson. Finally, it provides the coaches and teachers with a common focus for the observation. Below are some general questions that coaches might use in the pre-conference (Appendix 5):

♦ What math topic will you and your students be working on in this lesson? What are the big mathematics ideas in this unit?

♦ What do students need to know and be able to do to be successful with this lesson? How will you know if and when they are?

♦ What misconception might students have about the content you're teaching? How will you support student understanding? What questions and materials might help?

♦ How will you challenge students who are working above grade level and support students who are working below grade level?

♦ What would you like me to observe for?

Guidelines for Classroom Observations

Seeing is believing! Classroom visits are a great way to learn from one another, coach-to-teacher, teacher-to-teacher, and teacher-to-coach. In the classroom, the real application of what we know and can do becomes apparent. When coaches are being observed by teachers or teachers are being observed by coaches, the audience must be professional. In order to honor the classroom community, it's important to establish some guidelines for observing.

Classroom Observation Guidelines

♦ Come with a positive attitude and be a learner, not a critic

♦ Honor the structure and community of the classroom

♦ Be still and quiet—no sidebar conversations among the adults

♦ Maintain the focus of the observation, take notes, generate questions, catch student and teacher dialogue, read the walls

♦ Don't teach—remember you are a visitor in the room

♦ Bring your notes to the post-conference, with additional ideas and questions

Using Observation Forms

Classroom visitors may take notes about the focus topic in a variety of ways. Some like to take scripted notes on a blank piece of paper. These notes may look like a t-chart with two columns, one for what the students did and what the teacher did, or just flow from top to bottom of the page, in sequential order. Scripts catch authentic student dialogue and teacher questioning, and require the recorder to write as much as possible, as quickly as possible, including sketches and notation of mathematical thinking.

Each observation record looks different as a result of what the writer captures during the lesson. Later, the observer reflects on what evidence was present during the lesson that aligns with the purpose and professional goals outlined in the pre-conference and organizes the notes to present to the teacher.

Observation forms can also be used to capture the essential elements of a standards-based math lesson. Teachers, coaches, and principals commonly agree on what will be looked for ahead of time. The visitors intentionally record what happens during the lesson, under each category. For example, if the teaching goals for the school year include using cooperative groups, the think-pair-share strategy, rich mathematical language, and higher-level questioning, the observation form should reflect those ideas (Appendices 6 and 7). The visitors record specific examples of when and how these strategies played out in the lesson. If the teacher (being observed) has a special request to include or exclude some element from the lesson, that request is honored by the observers. In some cases, observers are also recording on-task behavior or gender equity during the visit.

When all teachers in the building are familiar with the categories on the form, they know what they and their students are expected to do during the lesson without any hidden agendas. The expectations are the same for everyone in the building, and this consistency is imperative for teachers who want observations to "be fair." When math coaches are observing and when they are being observed, these forms are used to record what happens during the lessons. Coaches will use these notes to celebrate success and plan for future coaching opportunities.

After several visits, coaches and principals become tuned into those essential instructional strategies and look for trends in classroom practice in one room, across grade levels, and vertically in the building. The specific data collected from many observation forms can be synthesized and shared anonymously with the staff, as in the example below:

> *Students and teachers are using lots of rich mathematical language in the classrooms this year. I heard kindergartners talking about a rhombus and third graders talking about the unlikely chance that green would be chosen from a bag drawing. This is so exciting! Most of the documentation I have captured from the forms points to an emphasis on teacher-to-student dialogue. Next quarter, let's focus on student-to-student discussions more. How can we make that happen?*

Planning For the Post-Conference

The pre-conference helps focus on the content of a lesson and how it's delivered. The class session allows the coaches and teachers to collect data; evidence of what occurred. However, the feedback provided during the post-conference is the most important part of the coaching sequence. Here, the coaches share information about what was actually seen, support reflections by using open-ended questions, elicit ideas from the person being coached, use data to set goals, analyze issues, and measure success. The coaches use nonjudgmental questions to probe for understanding; they paraphrase the teacher's response and use wait time to promote thoughtful reflection. In order to do this effectively, the coaches consider the observation data and determine the best way to share the information with the teachers.

For example, *"Did this lesson go the way you had planned? What were some of the highlights?"* Followed by statements like *"I recorded these open-ended questions...and noticed that you gave students some extended wait time to consider their ideas"* or *"Students were using parallel, perpendicular, intersecting lines, acute, obtuse and right angles, and all kinds of polygon names while describing their sorting schema. I heard some confusion about prisms and pyramids, and perhaps other three-dimensional polyhedrons. Did you see the same confusion?"*

The post-conference provides an opportunity to assess learning (for the teacher and students), inform instruction, and adjust educational plans. Effective coaches are able to focus on the connection between what should have happened and what did happen through careful questioning: *"How many of your students have at least one efficient way of determining the area of a triangle as a result of this lesson?"* Generally, the coaches take notes on what is discussed during this conference and what action steps may be taken as a result of the conference (Appendix 8). Math content coaches may help teachers plan their next lesson, identify key questions to ask to elicit student understanding, and/or gather additional support and challenge materials, as needed.

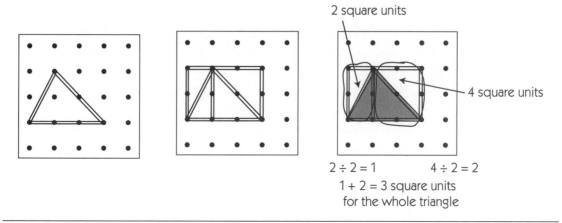

For example, student might build triangles and rectangles on geoboards to determine area, constructing a visual model for the traditional formula, (b x h) ÷ 2. Accelerated students may be interested in exploring other polygon shapes, using translations (flips, slides, and rotations) and documenting and sharing their solutions.

Post-conferencing is a great time to examine student thinking via authentic work samples. This is valuable to the teachers for several reasons. It increases the teacher's knowledge of his/her students. Even with a small class, a teacher can't hear or see all students' thinking during a lesson. This data about student understanding is used to inform instruction and tailor future lesson blueprints. It opens a discussion about the mathematics that students are learning and where gaps or misconceptions may lie. It provides a basis for discussions and reflections on the learning of each student in the classroom. Depending on the relationship between the teacher teams and the coaches' facilitation skills, the conversations around students' understandings and products helps teachers develop strategies to address those instructional needs. This is fundamental to good teaching and therefore, a powerful part of a coaching post-conference.

Getting Started On a One-To-One Collaboration

To establish a one-to-one alliance, coaches listen for the needs and wants of teachers, their content knowledge of mathematics, and implementation of standards- based pedagogy. The first step is to listen when teachers are talking. *"Will my students learn best if I…? Should I try…?"* At that point in time, the coaches offer a menu of options: providing resources, modeling, or collaborative team-teaching, with the ultimate goal of building the teachers own capacity. During the pre-conference, lessons, and post-conference, coaches mull over various perspectives and ideas. They collect evidence and data to further analyze the question and lead the teachers to a new understanding. Ideally, they continue this inquiry cycle throughout the year, with new *"I wonder…"* statements, gently nudging teachers to make changes in their practice that impact student achievement. This relationship is dependent on clear and honest communication, and trust and respect.

> *I was invited to a middle school classroom of a master teacher, with lots of academic and behavior challenges. We planned, team-taught, and debriefed about the big mathematical ideas and student proficiency with specific concepts and skills related to their grade level benchmarks. We also celebrated their engagement with the cognitively demanding task. After the lesson, the teacher said honestly, she couldn't imagine doing this with anyone that she didn't respect and trust. It takes time to establish that kind of rapport.*
>
> *Math coach reflection*

At each school site, coaches work with new teachers, experienced teachers, and teacher leaders. Each of these subgroups is in a distinctive place along a continuum of knowledge and experience. When presented with new information about mathematics content or the pedagogy of best practice, teachers require varying support from the coaches. Even accomplished teachers look for professional experiences to challenge them. As teachers personally experience these ideas with students and faculty, they apply or initiate these changes themselves. Over time, teachers develop automaticity with the skills and content and generalize this new learning to a variety of settings. Depending on the motivation and previous experiences of the teachers, this process may take weeks, months, or years to become a natural part of the teachers' schema.

This learning cycle is much like the reading process teachers use in the classrooms with students. Teachers "read aloud" text at a challenging level while they introduce students to an author, genre, or trait. Then they monitor additional practice in small guided group settings and provide specific feedback to the students about their acquisition of skills and comprehension. Finally, students take off and read independently, with fluency and comprehension. At this point students are able to decode new words and apply the syntax and semantic cueing strategies to a variety of texts with minimal support. Teachers continue to facilitate the reading cycle while they check for comprehension at students' instructional and independent levels. This gradual release of responsibility is part of adult learning, too. Math coaches present interesting mathematics in context, with support, give teachers feedback on their practice, and then continue to check in during independent work. Coaches are monitoring this process and adjusting to a wide range of adult needs constantly.

Everyone Participates!

Selecting teachers for one-to-one coaching can be a bit tricky. Preferably everyone at the school site is asked to participate in observing a model lesson, team-teaching a lesson, or being observed teaching a lesson, several times during the school year. This neutralizes the "stigma" of coaches being assigned to just ineffective or inexperienced teachers. It includes everyone and allows the math coaches to get started right away. Early on, principals and coaches might also establish criteria for choosing teachers for more intensive coaching cycles. *Will all the teachers participate this year? Are some teachers going to "have to" because they are new, new to a grade level, or identified as struggling with the content or implementation of a new program?*

Most school sites have a collection of accomplished, proficient, and basic teachers, and from time to time, teachers may switch roles because of personal and professional challenges. Accomplished literacy teachers may or may not be skilled math teachers. Proficient teachers under a good deal of stress may be working at a basic level this school year. Positive and hard-working teachers may have a reputation for being great teachers because people like them, when in fact they lack math content knowledge. Math coaches will want to investigate these staff dynamics and defer judgment until they have met with the teachers and worked in their classrooms. There are advantages in working with teachers at all performance levels.

If coaches begin with teachers who are identified as "accomplished," their journey will most likely be swift, successful, and public. These pioneers are teacher leaders and will further promote the mission of increasing student achievement through best practice. Their positive energy is infectious and proficient teachers at the school will want to be a part of that esteemed group. When coaches begin with proficient candidates, their journey is not as swift, but by and large successful. Helping good teachers understand the mathematics content more thoroughly and implement a standards-based approach helps these teachers do their job better. With the coaches' sustained support in the classroom and whole and small group workshops, teachers will witness an increase in student engagement and academic success and shift their practice accordingly. They usually need additional opportunities to see lessons modeled and more specific feedback on strategies they are practicing to become skillful. These teachers represent the largest group of teachers at most school sites. Once they catch on, the gains they make thrill and astonish even seasoned math content coaches.

Principals might also identify basic teachers who are low-functioning in the classroom to participate with the coaches on a regular basis. These teachers may not be ready or willing to change their practice or put any time and energy into becoming better teachers, regardless of the content area. Although math content coaches are frequently asked to work with this group of teachers, they should only do so with the principals' participation, since coaches are not supervisors. These teachers have the potential of drawing down the coaches' positive energy and monopolizing the coaches' time. They may be slow to progress because they either don't have the capacity or chose not to change. They might make negative experiences public and/or criticize ongoing coaching efforts as well as any new teaching ideas. Coaches count on other teachers to help engage them during whole and small group collaborative work.

If the long-term goal is to create a professional learning community, coaches will want to focus most of their resources on those teachers who are willing and able the ones that have the most potential to make a change, without ignoring the challenges. Before the first staff meeting, principals and coaches need to clarify how teachers will be selected for the more intensive one-to-one coaching cycles.

Planning for Coaching Cycles

Each coaching cycle is tailor-made to the situation. It depends on the school, the principals and coaches, the teachers, their coaching relationships, and the classroom of students. It is grounded in inquiry and reflection and is participant-driven. The coaching cycle is sustained, intensive, and supported by modeling and collaborative problem solving. Once candidates have been selected, coaches and teachers make a plan based on the inquiry topic and the professional goals and responsibilities of the participants. This plan is a road map of sorts, but certainly is revised to accommodate the situation. Active listening and objective documentation are the two characteristics of an effective coaching cycle. Math content coaches are classroom teachers who understand the importance of monitoring and adjusting their teaching with students. Adult learners respond to setting objectives, providing feedback, and reinforcing effort as well.

Coaching cycles may last five to 20 sessions, depending on the focus of the intervention. An instructional goal is identified by the coach and teacher, based on math content and instructional pedagogy. In the example below, 15 one-hour sessions are scheduled with the teacher and coach and specific actions steps are articulated.

♦ B. K. will plan and implement a standards-based unit on area, perimeter and volume concepts. This unit will include lessons using two- and three- dimensional shapes and solids and cooperative group work and be assessed by a performance task.

♦ Prior to this unit, B. K. and the coach will observe in a middle school math class to see how these concepts and skills are applied in a problem-solving context, and cooperative groups are used to engage students in the learning (one session).

♦ B.K. and the coach will read the unit introduction in the curriculum materials and the chapter on Measurement from the resource book in the library and outline the major landmarks in the development of these concepts. These outlines will be discussed at collaboration time (one session).

♦ The coach will initially plan the unit, and then revise it with B. K. (two sessions).

♦ The coach will introduce the first inquiry session, teach three additional sessions, and observe B. K. five times during the course of the unit (nine sessions).

♦ Towards the end of the unit, the coach will meet with B. K. to look at student work from the assessment task to determine student understanding of the concepts of area, perimeter, and volume (one session).

♦ At the end of the coaching cycle, B. K. and the coach will meet again to evaluate the success of the experience (one session).

Final Thoughts

As teachers are making changes, their self-confidence may decline. It doesn't feel good to "not know," to be in a state of disequilibrium. This is especially true during the first year of a new curriculum implementation or when teachers find themselves at a new grade level. Math coaches and building leadership need to support these individuals even more during this period of instability. Without question, the one-on-one coaching relationship is further enhanced by a positive schoolwide learning community and sustained by collaborative small group work.

A one-to-one collaborative setting provides customized professional development in a classroom context, at the teachers' stage of development. The job of content coaches is to find out what is interesting to the teacher, to gauge their comfort with the inquiry process, and to determine when to initiate the process. Regardless of the kind of one-to-one relationship teachers and coaches engage in, coaches document who they worked with, on what topic or lesson, and what the outcomes were of the interaction on a regular basis (Appendix 9).

5

Understanding Group Work

*"Dependence is the paradigm of **you;** you take care of me; you come through for me;*
you didn't come through; I blame you for the results.
*Independence is the paradigm of **I***
I can do it; I am responsible; I am self-reliant; I can choose.
*Interdependence is the paradigm of **we;** we can do it; we can cooperate;*
we can combine our talents and abilities and create something great together."

Stephen Covey

Combining the talents and abilities of the building teachers and math coach to maximize the teaching and learning for all students takes a tremendous amount of cooperation. It cannot be done by one person, or by individuals working in isolation. Math coaches who have been classroom teachers might recall how much time and energy it took to get a class of students to work as a community perhaps four to six weeks of daily interactions. Once coaching roles and responsibilities have been established, and goals set for each marking period during the year, it's time for coaches to begin extensive collaborative group work. *How exactly does this happen?* This chapter will take a look at group dynamics, working with teacher leaders and resistors, and suggest steps for building a "we" community.

Developing Collaborative Teams

School-based content coaches are often caught off guard the first time they present to or facilitate a group gathering. As a content coach they are now attending to mathematics content and teaching practices, while also trying to develop a culture of collaboration in group settings. *How do groups of people become a team?* Bruce Tuckman (1972) observed group dynamics as he constructed a model for team development. When groups are forming, they come together and get to know one another, and members define their role and importance within the group. This is the "*you* and *I*" phase of the collaborative process. People introduce themselves and find a place to sit in the library or cafeteria. Initially, there is a dependence on the coach for guidance and direction. *What is the purpose of the session? What are the objectives?* Here's a sample scenario:

> *The math coach may ask participants to examine the NCTM Process Standards by highlighting the verbs implied under each of the Problem Solving, Reasoning and Proof, Communication, Connections, and Representation Standards (Appendix 1). After members have identified those verbs, the coach may ask the participants to share their reactions to the verbs with another teacher near them. Once participants have shared their thinking with one other person, the coach may ask a member to record the observations of the group onto a chart paper. The coach solicits ideas from several teachers in the whole group. (This is an example of the think-pair-share strategy). "We noticed that they were active verbs. It reminded me of Bloom's Taxonomy, and Norm Webb's work with Cognitive Demand. I sure didn't learn math that way! Kids are doing something with the math ideas."*

In the example above, the math coach is leading the task and inviting staff members to contribute, intentionally crafting an opportunity for everyone to take part. Participants are actively listening to each other in pairs and as a whole group, creating a common experience and a common language about the process standards and what they imply about our current instructional practices. Some staff members may choose to remain passive observers. The math coach as an instructional leader must be prepared to create interest and motivation for everyone in the group to get involved.

Perhaps the coach dumps out a variety of jar lids, tubes, and cans from circular wastebaskets. She gives each team of teachers a tape measure and asks them to measure the diameter (distance across) and circumference (distance around) of each item on their table. These measurements are recorded onto a data sheet. Then participants are asked to examine the ratio between these measurements (discover Pi, between 3.1 and 3.2).

Circles Measurement Data			
#	Diameter	Circumference	R
1			
2			
3			
4			
5			
6			
7			
8			

If there are some teachers interested in a bit more of a challenge, they can graph the relationship (the horizontal axis representing the diameter and the vertical axis representing the circumference) onto grid paper. The slope of these equivalent ratios should make a line close to 3.1. In this mathematics task, teachers are working at a variety of readiness levels, with the support of colleagues. Teams of teachers enthusiastically share out their learning with the whole group.

After the inquiry, the coach may request that participants evaluate the experience based on the active verbs implied by the process standards. *Did the task encourage them to investigate, analyze, use, apply, adapt, and communicate? Was it worthwhile? What did the experience feel like to them? How would students feel if they were asked to think about mathematics this way?* The coach poses these questions and encourages everyone to privately think about this, and then records their thinking on a slip of paper. Participants get up from their table and go visit with three other people in the room. They communicate their thinking and actively listen to a range of responses from people at other tables. This is sometimes referred to as a "give one-get one" strategy. A summary of the explicit strategies used by the math coaches at the end of the professional development makes the purpose and objectives clear. It also creates a shared experience for future group work.

At some point in time, there is a difference of opinion or a decision to vote on, and then the group starts **storming.** This is a conflict stage when diverse styles and goals emerge and people express their frustration: They don't understand the research, think that the administration just wants change for change sake, or think others talk too much. Since math coaches want to promote harmony, this stage is generally uncomfortable for them. However, coaches must honor this stage as part of the process of group development because this is where the work of "we" first begins. Participants need to see themselves as individuals with something to contribute and work through their disagreements to form a bond, united for a common good. An investment is made in the group through this conflict. Skilled coaches provide a safe climate for everyone to contribute in shaping that perspective. The following is an example of a storming stage, in a group task:

> *In many schools around the country, debates about mastering the basic facts and developing computational fluency immediately initiate this storming stage. Math coaches may choose to have participants read a collection of current articles from the NCTM journals about computational fluency prior to the group meeting. Once there, the members get into jigsaw groups. Members that read article A, get into smaller groups and summarize the big ideas in the article, becoming the experts on the content. In a smaller group, more people get an opportunity to share their thoughts and experiences. Other staff members that read Articles B and C are doing the same. After a period of time, the math coaches ask participants to put themselves into ABC groups of three and summarize the articles for each other. What are the likenesses and differences in the articles? What do the staff members agree on and what do they still have concerns about?*

When the articles have been read and reviewed, the whole group gathers together, and members share their summaries, agreements, and concerns. The math coaches are now facilitating and monitoring the discussions among the group members, no longer directing the task. Everyone's ideas are listened to respectfully, even if other members don't agree. Notes are recorded for further discussions by someone other than the coach so that the coach is free to listen attentively. The coaches may suggest that the group determine the next steps. In some situations, consensus is better than voting (yes/no), because it's an agreement everyone can live with. The topic has been discussed, and then perhaps the group reaches a compromise about practicing fact-retrieval strategies prior to the implementation of random drill sheets. This leaves the door open to discuss, *"What do we still wonder about? Who might be interested in doing some additional reading and research on the topic?"* The responsibility begins to shift onto other group members, and not just back onto the coaches.

After a period of time, the group works out its own set of expectations about how they will operate, challenge each other, and support each other's ideas in and away from the meetings. This is the **norming** phase where the group establishes the "we" protocols. The participants must do the forming and storming work in order to move forward on their journey. At this stage, people are vested in a relationship with one another and recognize that collaboration is a better way to support the group's vision. Everyone is encouraged to participate and contribute, and they are given multiple avenues for doing so. Math coaches continue to actively recognize the contributions of individuals while emphasizing the influence of the group. The experience below is an example of a group getting to the norming phase. The math coach facilitated this group through this development by focusing on the shared vision: How can second graders show us what they know and can do in mathematics?

> *A team of second grade teachers met with the coach to determine what assessments would be used at the district level to judge student mastery of math concepts and skills. Initially, teachers wanted a multiple-choice test that would be easy to score. Several members of the team attended an assessment workshop and recommended some constructed response journals prompts and a performance task. After meeting for four months, the team had a product, a collection of multiple-choice items, journal prompts, and a performance task with a scoring rubric. Members worked through their professional differences, teachers contributed to the project, and as a result, they have bought into the assessment plan.*
>
> *Math coach reflection*

The fourth stage is where the group gets the job done—they are **performing** and feeling really good about the work. This congenial atmosphere, where people are on equal terms, makes the work fun and rewarding. The group fellowship is like a family. If someone new comes into the group at this point, they may notice an alliance among the group members that doesn't include them in the inner circle. Occasionally, group members act independently of the math coaches. Although this may feel threatening to the coaches, it is evidence that the participants have embraced the work themselves. They don't need the coaches there to supervise the work! They own it.

> *Several elementary teachers wanted to examine interventions for at-risk students. They got together after school one day and started looking at the research and the supplemental math materials that were available on line. They asked the math coach to join them the next week so they might present their findings and take a look at implementing some of these interventions schoolwide. They even identified their next step—to find appropriate assessments to identify their students and monitor their progress.*
>
> *Math coach reflection*

This teacher team has a high degree of autonomy and efficacy. They may check in with the math coaches to be sure their work is aligned with the goals of the school and their professional learning community. Members in this group appreciate recognition for a job well done and warrant the support of the administration and coaches.

Some groups will eventually come to an end because the task is done, grade level assignment changes, the course is over, or someone moves from the community. At that point in time, the group **adjourns.** To bring closure to the group relationships, math coaches may organize a celebration and share the history of the group, including artifacts that honor each member's contribution and the final product.

I was part of a cohort book study group. We were the pioneers and needed each other to continue the work we were doing in our classrooms and schools. We had class after school from 4–7, and even at 7:00 in the evening, we stood in the parking lot with the wind howling, for just another minute of "we." Two years later, we are still meeting for supper because we so liked each other personally and professionally. We never want to adjourn.

Math cohort member

Tuckman is quick to point out that these stages are not linear, but rather dynamic and cyclical. As new members or new tasks present themselves, the group will shift its focus and reorganize. Coaches must be constantly aware of where their participants are in the process of collaborating and what their individual and group needs are as a result of that stage. As facilitators, math coaches must be willing to work through conflict and release the responsibilities to the group members.

Creating Agendas

When creating agendas for scheduled meetings, coaches focus on the substantive items related to the group's overall mission and do not let little odds and ends tie up the meeting time. Memos can be used to communicate information and free up valuable time for collaborative group work. The agenda should describe the action (verb) and tasks to be completed. For example, "The cohort will examine algebraic tasks for standards and benchmark alignment and potential for student engagement. Be sure to bring an example of a worthwhile grade level math task to examine." Some agenda items may include a time frame to ensure the pacing is reasonable. If the item is a "how do you feel about using calculators for this unit," and coaches want teachers to reflect on the idea privately, they may ask them to journal about it and be ready to share out next time the group meets. Coaches may suggest participants prepare for a future meeting by gathering student work, assessment data, reading an article, unit introduction, or reading a series of lessons in the unit. This fosters intradependence among group members and shared responsibility. It continues to encourage participation from everyone. Members may also contribute topics for the future agenda based on the overall goals of the group. Agendas can be sent out via e-mail or notes put in the office mailboxes. As a general rule, coaches plan for one or two main items for each hour of allocated meeting time.

Running Meetings Effectively

During meetings, coaches ask one of the other participants to be responsible for taking notes on the discussions, questions, and group decisions. This record will help create a record of the groups' performance. *What did we do? What did we decide? Who will get it done? When will we meet again?* It can also help clarify something that was said and perhaps misinterpreted by others. This frees the coaches up to facilitate the discussions without worrying about getting everything recorded properly. It also shares the responsibility for running the meeting with other group members.

Most organizations operate their meetings based on some professional ground rules that include beginning and ending the meeting on time, keeping side conversations to a minimum, working towards consensus by actively listening to each other's ideas, and supporting the final decision of the group. It is a good idea to have participating staff generate a list of agreements that are important to them as a professional community. Although this takes a bit of time in the beginning of the year, it sets the stage for future interactions and supports the *norming* phase of collaborative group work. The group recorder can take notes on what items are mentioned by the group and publish them for everyone to refer to during the school year. Any items that were omitted from this initial list can be added to the list later. Below is a sample list of Group Agreements:

Group Agreements

♦ Listen actively, and try to understand the other person's position

♦ Disagree respectfully and use evidence to support your ideas

♦ Work on the work—no side projects

♦ Stay on topic

♦ Stay focused on the students. What's good for them may not be easy for us,

♦ Share your ideas and solutions, your struggles and successes

♦ Turn off cell phones or turn them on vibrate if you have an emergency

Math coaches must prepare for each agenda item and set a clear goal for the session. Posing a question or problem for the group to investigate during the meeting is one way to write a goal for the meeting. For example, *"How do students develop place value concepts?"* Prior to the meeting, coaches read current text on the topic of place value understanding. They ask teachers to read their place value unit introduction from the current curriculum materials before the meeting. When they arrive, they begin to discuss what they read in pair groups. While they are sharing, coaches ask teachers to consider what concepts and skills are necessary for students to "understand place value." *How would they know if students "got it"? What experiences have students had with models and manipulatives prior to this unit?* Then, as a whole group, coaches solicit comments supported by evidence from the reading. Coaches practice active listening and paraphrase participant's point of view, without bias. They anticipate some conflict and tension during the session, and keep in mind that teachers will need to challenge their own assumptions to take hold of new learning. Perhaps the coaches use student work samples from a variety of stages in this development and ask participants to evaluate them based on the continuum of learning about place value concepts. *Which students have a conceptual understanding of ten and ones? Which students are still likely to count up by ones because they are not able to use grouping strategies? Which children are able to decompose numbers is flexible ways?*

Roberto's Way	Emma's Way	Lucy's Way	Travis' Way	Midori's Way
$36 + 28$ $30 + 20 = 50 \quad 6 + 8 = 14$ $50 + 14 = 64$	$20 + 30 = 50$ $50 + 8 = 58$ $59, 60, 61, 62, 63, 64$	$28 + 36$ $28 + 2 = 30$ $30 + 34 = 64$	$28 + 36$ $28 + ②= 30$ $30 + 36 = 66$ $66 ②= 64$	$^{1}28$ $+ \underline{36}$ 64

Students may solve two-digit addition problems with regrouping in a variety of ways. They might split the tens and ones (partial sums), decompose or transform sets to determine the sum.

While doing so, coaches continue to treat all participants with respect, and foster a safe environment, even when someone challenges them. As a group, they begin to answer these questions: *How do students develop place value concepts? What are the concepts and skills that need to be mastered for students to be successful? When and where are these concepts taught and mastered? What manipulatives and visual models might help support this development?*

Working with Other Teacher Leaders

Establishing relationships with other teacher leaders in the school community will be essential to the group's success. Some teachers positively impact others by encouraging, mentoring, sharing, and staying focused on the positive aspects of the work with students and their families. Other teachers lead in a negative direction, with sarcasm and references to other failed attempts at making a difference. Math coaches will want to think about how these positive forces can be mobilized and how the negative forces can be minimized.

Coaches might identify teacher leaders who love to learn and initiate them as part of the support system. These leaders create a base for decision making and moving the school vision forward. Possible leaders may have completed their National Board Teaching Certification, are involved in community or professional organizations, are award winning teachers, or are math department chairs. Perhaps they are pursuing leadership opportunities on their own, like district math curriculum work or assessment development. These teachers are a positive conduit between the students and learning and a vital part of the professional community. They already devote their energy, time, and expertise to the collaborative process, offering others feedback and resources, and are most likely to cherish the idea of being asked by the math coaches to contribute to the school vision.

Dealing with Resistance to Change

Math coaching is about change: Change in instruction, change in curriculum materials, assessment practices, meeting the needs of students, and meeting the academic requirements of No Child Left Behind. Not all change is good and ultimately the changes coaches suggest must satisfy the needs and values of those whose support is essential. Overcoming resistance to change is not the same as creating a commitment to change. Some staff members will choose to remain passive during group work, and not challenge the coaches' ideas. Although this seems like it's a good thing, coaches need to keep in mind these people will not commit to making a change because they are observers, not participants. Coaches are asking group members to exchange their time, energy, support, creativity, and insight for something of value to them. The first objective will be to identify *what* they value, so that coaches can address their needs and wants. Phillip Schlectly (1997) suggests considering several questions when "marketing a change effort." Reflecting on these questions with teacher leaders, administration, and other content coaches early in the year will save coaches lots of frustration down the road.

Thinking about Change

♦ If this change were to be implemented, what are the critical values held by the constituents who would be affected?

♦ Are these values likely to be manifest in different ways by different groups or subgroups? Which of these values are likely to be served by the changes and which will be threatened by the change?

♦ How can the proposed change be organized and implemented so that the values served are increased and the values threatened are minimized?

♦ Can these modifications be made without threatening the integrity of the change?

♦ If so, why not make them?

♦ If not, is the proposed change possible at the present time or will the lack of support eliminate the possibility of success?

♦ If the change is not possible, what might be done to prepare the situation for the change?

Not all group members will need or want the same things at the same time and coaches will need to differentiate the meeting and learning environment to accommodate the multiple values at play. The *Concerns-Based Adoption Model*, sometimes referred to as CBAM (Loucks-Horsley, 1998), is a framework for looking at the "Stages of Concern" in learning and development during which a person's focus or concern shifts in fairly predictable ways. During the three early stages, Awareness, Informational, and Personal phases, individuals focus on themselves: *How does this innovation change my work?* Midway through the seven stages, the Management phase focuses on mastering the routines: *How can I manage the time and the materials?* The Consequence, Collaboration, and Refocusing Stages are focused on the results and impact of the curricula and practices: *How does this affect my students?*

In-service geared towards the "how-to" will most likely meet teachers at the first three lower levels. This unfortunately is what has been offered in past professional development and has not proven to be effective in changing teacher practice. Math content coaches can help teachers move through the Management level the first year, towards the Collaboration level of practice where they improve their own performance and thereby increase student achievement. This process generally takes two to three years, and varies based on the teacher's attitude towards the innovation. If ongoing support is not available, teachers are likely to feel overwhelmed and frustrated, discard the innovation, or adopt poor practices. When a teacher's concerns are met at the approximate stage of their development, they are likely to move towards the next stage.

This is why math content coaches will want to create additional opportunities for small group work that are based on readiness and interests. The small groups can be more flexible in their approach and delivery, and continue to sustain the work of the whole group, while meeting the personal requests of individuals.

Monitoring Group Work

All teachers need recognition and affirmation for what they have done well. Change initiatives that elevate teachers' feelings of worth rather than denigrating them for not doing enough, have a better chance of surviving the first stage of group work. Coaches help form group teams by recognizing that each teacher is important to the future implementation of the proposed professional learning community or change initiative. This affirms not only the person but the goal itself. Coaches may review the vision statement regularly to elevate the worth of the organization and state this declaration as if it is already happening.

Most math coaches are supporting veteran staff members along with novice teachers. These groups will think differently, behave differently, and feel differently about the profession. When introducing a topic in a whole group setting, coaches honor the current research and theory along with the personal experiences of the novice and veterans. Some of these veterans will want to do what they have been doing; even if what they are doing is not successful, they know *how* to do it. They have files of math worksheets run off for extra practice when their students don't "get it." These veterans may already anticipate that their students won't experience success in the way they have taught, but will continue to hold fast to the "way we do things around here." Breaking patterns of behaviors and changing values is no small matter. Novice teachers can feel unsure about taking risks as well. They don't have the personal and professional experiences to draw upon when an initiative is presented. When that change disrupts their routines, it creates more uncertainly. This uncertainty is based on fear. Overcoming this fear through collaborative culture is possible. As Eleanor Roosevelt said, *"You must do the things you think you cannot do."* Moving from the "you" and "I" to the "we" increases the potential of creating something great together, and decreases the fear factor. Coaches encourage teachers to do the things they cannot do, together.

Choosing a Fractal Experience

When school leadership begins by selecting a small fractal experience from the school improvement goal that can be easily monitored by a pre- and post-measurement, followed by a systematic implementation and summary of its impact, they are more likely to move the success of a larger-scale change initiative forward (Galvin, 2007). For example, comprehensively implementing a standards-based math curriculum is a large-scale initiative for many schools. A fractal experience that contains the same elements as the larger proposal and establishes a pattern of change could be connected to the use of the NCTM Communication process standard. In this standard students are encouraged *"to communicate their mathematical thinking coherently and clearly to peers, teachers, and others and use the language of mathematics to express mathematical ideas precisely."* The fractal experience may include the following steps: Key vocabulary words are identified by grade level teams for the first unit of study. Word Resource cards, with non-linguistic representations and definitions, are introduced in the context of a lesson and then posted in the classrooms on a math word wall. See Word Resource Card on page 64.

The teachers "couple" informal student talk with formal math language during the lessons. Desks are grouped in pods of two to four students to encourage oral communication during the math investigations. Walk-through classrooms observations are used to monitor classrooms for the use of coherent and clear communication, and written student work is collected at the beginning and end of a unit of study. Teachers then review the impact of vocabulary and cooperative group structures on student communication skills, as demonstrated by the walk-through observations and student work. Math coaches facilitate the review of the fractal experience, attributing student communication gains to the collective effort. The staff then identifies the groups' next steps. This establishes teacher efficacy and collaboration as a professional learning community and sustains school improvement efforts.

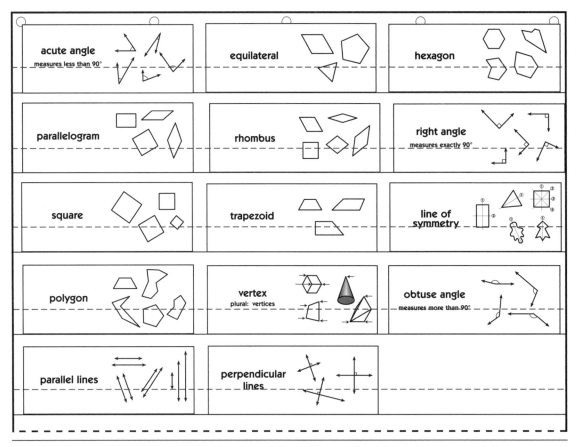

Word Resource Cards can help develop rich mathematical language especially when graphic representations (pictures) are attached to the vocabulary word. These cards also have a student-friendly definition on the back. Teachers can post them in pocket charts during the unit of study and then move them to a math word wall. It is recommended that these words are grouped by strand. For example, all the geometry words would be together, angle words together, two-dimensional shapes together, and so on. English language learners benefit from this explicit vocabulary support.

Implementing something new and getting good at it, will take time and energy. This loss of time, energy, and success is what people fear the most. Keeping an eye on short-term goals will help teachers and administrators see the results they need to keep the fear factor from unraveling the innovation. Math coaches rally and focus on how *we* make a difference, and how this new intervention will make a difference as well.

Some coaches begin or end their meetings with a journal prompt. For example: *"How will using a think-pair-share structure help all your students communicate mathematically? Share some examples and concerns."* In this way, coaches communicate the value of impacting students in a positive way and validate a range of classroom experiences. Time is set aside during the next meeting for addressing concerns such as the amount of time the strategy will take up, how to deal with special needs students, or how to handle student groups that are not working well together. When teachers share their fears and struggles in a safe environment, they are more likely to resolve them. If this sharing doesn't occur, many teachers hold on to their resistance and don't let go.

Coaches continue to foster collegial support for each member of the group and support for special interest groups (like novices, veterans, math-phobic teachers) and maintain the focus on the whole community. Other positive teacher leaders in the group can provide the coaches with necessary feedback about what they see and hear during group meetings and small group breakouts. Coaches use this feedback to determine where the teachers are in their group development, mathematics content knowledge, and understanding of best practice strategies.

Change requires a commitment of energy and resources. It requires people to take risks and break habits. It causes discomfort and uncertainly. It creates needs as well as satisfies them. In times of change, where stress is high and security low, these values (needs and wants) will dominate the group development.

Suggestions for Building a "WE" Community

♦ Ensure that policies and procedures support collaboration by sitting people around tables and chairs rather than an auditorium-style lecture hall, and create specific time during your group meeting for collaboration

♦ Identify and support teacher leaders who will support the effort to make changes that focus on improving student learning, through curriculum, assessment, and instruction. Share facilitation roles and responsibilities with them when appropriate

♦ Use data to plan for changes. This data may come from the group, the district, or state assessments, and/or from parents and students. The data is not used for grading teachers or students

♦ Hear everyone's voice in the group (don't always act on what you hear)

♦ Continue to plan "fractal" experiences and celebrate short-term success

♦ Examine the teaching and learning community and its impact on students to determine the next steps

Final Thoughts

Whose support is needed to make group work productive? Everybody's! However, active support of some stakeholders is more critical than others. If coaches wait until everyone is ready, they will wait a long time. Coaches call first on those persons, groups, and agencies that are needed for active support—by changing the way they behave, by giving up traditional interests, or by providing additional resources. They develop efficacy with small scale initiatives and maximize the teaching and learning opportunities for all stakeholders by understanding people individually and collectively as a group.

6

Working with Groups

*"With regard to excellence, it is not enough to know,
but we must try to have and use it."*

Aristotle

Many professional development experiences are about "getting knowledge." Teachers file into the library, sit down, take notes, and leave with new information that they set aside to get back to the business of teaching. Aristotle understood the importance of implementation, the importance of using the new information towards excellence.

When schools create opportunities for teams of teachers to process knowledge through collaboration, they enhance practice, teacher satisfaction, and student learning. Investing in math coaching and quality staff development is one way of sustaining ongoing collaborative adult learning and ensuring the implementation of new ideas long after the workshop in the library is over.

This chapter suggests ways coaches might facilitate large groups of building staff (up to 30) during a professional development day or curriculum training when the information is appropriate for everyone, and small groups of three to eight people, perhaps grade level teams looking at student work, curriculum units, and assessment targets. Some coaches may also be asked to facilitate standards or curriculum training, essential learning, or assessment alignment committees. They will need to determine the priorities for their building and plan for the group work accordingly. Key questions to consider are: *What is the purpose of the group work? Who should attend? Is the group work a choice or required by the administration? Would the size of the group affect the participation and collaboration? What is the identified product or expected outcome of the group work? How much time will be necessary?*

Presenting to the Whole Staff

Before the school year begins, math coaches are sometimes asked to present a buildingwide math staff development session. After they have reviewed the curriculum materials and discussed their first steps with the principal, math coaches will want to identify and prepare several rich mathematical tasks and plan to highlight the researched-based, instructional strategies they will be using during each of the staff development sessions. In this way, they are not only increasing the math content knowledge of the teachers, but modeling what classroom practice should look like, sound like, and feel like, directly for the participants. This teaching practice should also be a reflection of school and district goals and clearly articulated by the administration.

If the math coaches are new to the staff, they'll introduce themselves briefly and share their enthusiasm for the role they'll play in the community of professionals at the site. Returning coaches may want to share a brief personal experience and jump into doing mathematics. Cartoons, music, picture books, poems, or props can also be used to "capture" the audience's attention. Then coaches "captivate" the audience with a problem-solving task and give teachers private work time with manipulatives and visual models. These manipulatives and visual models (two-dimensional drawings or sketches) are the nonlinguistic representations that help create physical models and mental pictures. Teachers and students will use these visuals to make sense of the mathematics in the task.

This activity incorporates both Tangrams and fractions. The Tangram puzzle was invented by the ancient Chinese hundreds of years ago. It is a square cut into seven pieces. When rearranged, these pieces form a variety of shapes and pictures. This puzzle provides an excellent background for determining fractional parts and wholes. Each pair of teachers will need a set of Tangrams and some journal paper. Overhead and magnetic Tangrams are also commercially available.

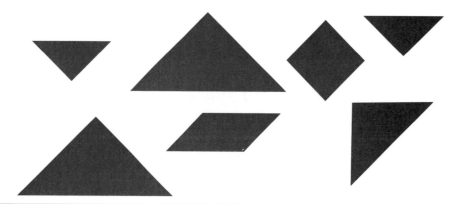

Ask teachers to build a square unit from all seven pieces. If the square is one unit, invite participants to determine the fractional part of the whole for each piece. Invite them to represent their thinking, a proof, of their work.

Large right triangles: ¼, *Small right triangles:* ¹⁄₁₆ each *Small square,*
Medium Right Triangle and Parallelogram: ⅛,

For additional challenge, change the value of the unit. For example, if a large right triangle is one area unit, what is the fractional relationship of the other pieces?

Square, Medium Right Triangle and Parallelogram: ½
Small Right Triangle: ¼ (each)

During the sustained work time, math coaches resist telling participants what procedure to use and how to think about the problem solving task. They model extensive wait time and use open-ended, higher-level questions like *"What do you notice? How did you think about it? Can you tell me more? Would that always work? Why? What other connections can you make to fractions?"* Some coaches use a cue card with these questions on it to assist them in the monitoring process. If individual teachers finish quickly, coaches may ask them to solve the same task using another strategy or area model.

Next, math coaches have teachers share their work with a partner, or group of four participants. Finally, coaches invite at least three teams to share their strategies, using the visual models, with the whole group. This encourages participants to create meaning for themselves by making specific mathematical connections, asking each other questions, visualizing, inferring, and predicting and monitoring their own thinking (Hyde, 2006). In a genuine problem-solving task, there are a number of different approaches to solving the problem, even when there is one correct answer. Math coaches encourage discussion among the staff, and let the learning emerge from the participants, "uncovering" the mathematics through the experience. Processing this information is an opportunity to synthesize the learning for everyone, thereby deepening an understanding of the math content. There are additional Tangram lessons online that involve cutting apart personal Tangram sets from a paper square, area and angle measurement lessons, and lessons involving translations (flips, slides, and rotations). Perhaps some teachers only used Tangrams as a puzzle activity in the past.

Finally, math coaches may facilitate a debriefing about how this way of teaching and learning is similar or different from what teachers experienced as students in a math class. *In their math experience, did the teacher call on a few students who had their hands raised? Were they allowed to investigate a concept with a partner? Did the teacher stand at the front of the room and explain a concept? Or did students come to the board to share their ideas?* Participants may be asked to identify the instructional strategies used during this scenario (Appendix 4). This interactive instruction relies heavily on discussion and sharing among participants. Teachers, and their students, learn from their peers to organize their thoughts, develop rational arguments, and develop social skills as a community. Prior to this session, it is essential for the math coaches to carefully outline the math topic, allocate a reasonable amount of discussion time, organize the composition and size of the groups, and consider which reporting or sharing techniques will be used. Interactive teaching also requires the coaches, the teachers, and their students to increase their observation, listening, interpersonal, and intervention skills. If teachers are not using an interactive teaching approach in their classrooms, the group session will likely require additional clarification.

At the close of the session, participants are given some time to reflect on what they would like to take back to the classroom with them. *What did they learn? What can they apply to their teaching practice?* Math coaches ask pairs to talk about their learning and make time for a few members to share with the whole group. The time spent at the end of the session, identifying what participants are going to utilize in their classroom, is imperative for teachers to transfer these new skills.

Many coaches use standards-based mathematics lessons the teachers will be using with their students directly from their program materials or other quality resources available from the NCTM Web site or monthly journals. They are organized and well prepared for these lessons, the discussions, and the closing comments because first impressions matter.

Using PowerPoint

Extensive PowerPoint or series of overheads can give the impression that the coaches are the source of knowledge, or authority on the subject, and therefore should be used sparingly. Coaches using this style of presentation should limit the words on each slide to about 15 words, and use bold, large, and simple fonts. They should remember to use the slides to enhance their presentation, *not as the presentation*. No one wants to listen to someone read a series of slides out loud.

Staff development is enhanced by rich mathematics tasks and conversations among teachers. Successful coaches design and conduct professional development experiences that will have the potential to transform teachers' knowledge, values, and classroom habits by making direct connections with the practice of effective teaching. If the coaches lecture the group using a direct instruction model, teachers will be inclined to think that they can continue to do the same with their students. A short introduction that states the goals of the session, followed by hands-on math tasks and a summary of the teaching strategies used is generally more productive.

At the end of the session, coaches may ask teachers to give them some feedback about the math tasks, the delivery and facilitation of the instruction, and their individual needs for the school year on a form (Appendix 10). In this way, they have opened the door to communicating about teaching and learning and taken a risk. As frightening as that is, coaches will be asking teachers to do the same with them in the coming weeks. When coaches emphasize their role as a learner, genuinely interested in mathematics and improving instruction, many teachers respond in kind. When we *"try and use knowledge"* we come to know excellence.

Depending on the purpose for the whole group meeting, and time of year, there may also be some administrative business to share about assessments, standards and benchmarks, and curriculum materials. Coaches will want to be aware of these requirements and schedule their time appropriately.

Suggestions for Whole Group Collaboration During the Year

When planning for whole group professional development throughout the year, math coaches consider the needs of the school community and the time allocated. *What short-term goals need attention? What instructional strategies are currently being implemented? Which concepts are still challenging for students to learn? What is the focus topic for the year?* For example, if vocabulary development or writing in mathematics is a key topic for this school site, *how can the coach facilitate whole group sessions that increase these opportunities for teachers? How can teacher and student work be used to show notation, representation, and reasoning and proof? How can these experiences be presented in the context of writing and using clear and precise math language?*

When possible, math coaches choose interesting, mathematical tasks that are presented in unique problem-solving situations. They practice demonstrating and facilitating the group lessons, using instructional strategies that will increase student achievement. They ask participants to reflect on their instructional delivery and illuminate their teaching decisions for the teachers. Coaches can't assume that the teachers notice the use of higher-level questions and wait time. They invite teachers to try out some of these strategies in their classroom and share their successes with the group members at the next meeting. Ultimately, teachers must understand the mathematics concepts that they teach and learn new ways of enhancing communication in the classroom environment or they will continue to teach the way they were taught. Over time, as teachers experience standards-based mathematics as learners, they come to appreciate how tasks can be solved in a variety of ways and allow for multiple entry levels in their classrooms.

Coaches sincerely compliment the group for taking risks and challenging their instructional practice. They take advantage of the group's inherent need for collegiality and collaboration by building in some time for reflection at the end of each of the whole group sessions. Math coaches may take notes in their journal or logs about teachers developing new understandings with math content, how they are making connections to other topics and gaining confidence in their ability to do mathematics. Monitoring and reflecting on the established professional goals, math coaches adjust their staff development presentations to accommodate the needs of their audience.

Implementing a New Curriculum

In some situations, math coaches will be asked to help facilitate an adoption and implementation of a new math curriculum. Purchasing and unpacking the materials is just the beginning. When schools adopt a standards-based curricular program in its totality they are able to show an increase in student achievement over a sustained period of time. After all, these standards-based math curricula were written, field-tested, and followed rigorous research and design protocols, and the student achievement was documented when the materials were implemented comprehensively (Fullan, 2001). In this case, the responsibility of the math coach is to get to know the curriculum materials at multiple grade levels as quickly as possible. This might be done by reading overview materials provided by the publisher and unit introductions. As the year progresses, math coaches get to know the curriculum in depth by modeling lessons and observing in the classrooms.

Some schools adopt parts of the math program that complement their existing materials, and let teachers dabble in both worlds. These schools may see a short-term increase in student achievement due to the renewed interest in something novel, but when things get uncomfortable, teachers go back to what is familiar. This level of adoption is parallel to the concept of first-order change where the focus is placed on improving the efficiency and effectiveness of the existing system (Marzano, 2005). No big changes are made because the underlying assumptions about teachers' values and how students learn were never dealt with. In this case, new curriculum was purchased but the underlying problem about how the curriculum is presented has not been addressed. Teachers may use games and activities out of the program, but use them in a habituated manner. Unfortunately, these schools adopt some of the language of reform, tape posters in the hallways, and change nothing about their underlying assumptions. These schools are not invested in implementing the new curriculum materials if it requires a change in their practice (Fullan, 2003).

Math coaches and school leadership teams must initially consider the school climate. *Are their teachers currently adopting the new materials and resources? Is the staff implementing the materials with integrity? Do the teachers use the curricula materials in a direct instruction manner, not really changing the process of teaching and learning mathematics? Or is the shrink-wrap still on the materials in the corner of the rooms?* The solution to this implementation issue is complex and depends on the administrative support, the school goals, and the coaching and professional development resources available. *What are the nonnegotiable items teachers must do? Who will ensure they do those things? What is the role of the math coaches in this setting?*

The standards-based math teaching initiatives are directed at challenging the personal ideologies and values of the group members, which then leads to second-order change (Marzano, 2005). This level of implementation requires significant restructuring of the existing system. It is more successful in increasing student achievement and increasing teacher content and pedagogy knowledge in mathematics. It requires ongoing coaching and professional development for teachers because it asks them to consider the impact of their teaching practices on what students learn, and how they learn it, and how all students can be supported in learning significant mathematics. A shared responsibility for instructional improvement binds the teachers, coaches, and administration to making a change. This is the collaborative efficacy that promotes school improvement.

When the administration is not supportive of the changes necessary for a comprehensive implementation of a new standards-based program, coaches navigate a slippery slope. The coaches may be working with teachers at a variety of implementation levels, with no authority to move them along in the process. In this case, coaches will be working with the teachers that are willing and able to make the journey this year.

Planning Professional Development with Outside Consultants

Math coaches may also schedule ongoing professional development for their staff in the area of mathematics with outside consultants. In this case, they need to communicate clearly with the trainer, their expectations for high-quality professional development and insure that the content of the presentation is aligned with the school's goals in mathematics. As the host, math coaches may also be responsible for organizing the travel for the consultant; planning the day, time, and place; gathering necessary materials; and evaluating the professional development experience.

Working with Small Groups

Content coaches are concurrently charged with two major responsibilities: to increase student achievement by changing the instructional delivery and to create a collaborative learning community. In addition to large group settings, coaches will want to explore working in small groups as well. Coaching relationships flourish when they are based on trust and when the connection between colleagues feels mutually inclusive and supportive. The success of small group collaboration depends on the facilitation skills and tools that math coaches bring to the table and the group members. In a small group, everyone's voice needs to be heard. Coaches are working on listening without forcing their own opinions, so that the most reluctant members invest in the group work.

Coaches identify what kind of group they want to facilitate. *What are the needs/wants/values of the members?* They set guidelines about the best size for this group, how much time they need, when the members of this group can meet, and where they might be able to meet. For example, looking at student work is a worthwhile task, but trying to respond to 50 teachers at various grade levels could be challenging for a novice coach. Many coaches begin with small groups of three to eight people and increase their group size with experience. If the group meets outside the school day, teachers will decide which group they want to participate in. If the group meets during the school day, participation is generally mandatory.

Coaches may choose to facilitate all the meetings, or share the facilitation with another teacher leader. Some small groups rotate responsibility among all their members for planning, hosting, and facilitating the group. Depending on the experience and comfort level of the participants and the purpose of the work, coaches consider which facilitation model fits best. They establish the small group community norms and ask a recorder to document the conversations and decisions. If the meeting is at lunch, before or after school, they plan to serve refreshments at the first meeting and decide how the additional meetings will be catered. In most cases, meetings will run up to the minute people need to be at their next place of duty. If that is not the case, coaches set a finite time to end the meeting.

Some coaches like to collect feedback after each meeting. In this case, participants record any ideas or concerns they have on index cards or sticky notes, and coaches use these to discern patterns and anticipate the needs of individuals and the group. Open-ended questions encourage people to provide more than a yes/no answer. This specific and personal feedback helps monitor and adjust the group work to the needs of the participants. Consider using the following prompts: *What did you learn today? What will you do as a result of that learning? What would you like to learn more about?*

At the end of the year, or whenever the group is "done" with their work, coaches adjourn the group, and plan a celebration to highlight the work of individuals, showcase the product developed by the group, and document the journey taken. Photographs and anecdotal records can be shared at the commemoration.

Making Time for Small Group Work

Along with building leadership, coaches may be able to review and revise the school schedule to find time for small teams of teachers to meet during the day for at least an hour, once a week. Most often, this can be accomplished by scheduling specialist time (art, music, physical education, library, or computer lab). Some schools have arranged a schoolwide schedule where they send students home an hour and a half early on Wednesday, or begin later (with students) one day a week to make time for collaboration. This time may be shared among other content areas and teaching responsibilities (e.g., language arts, science, social studies, and technology).

Math coaches might make time to discuss key topics during faculty meetings, follow up during the scheduled grade level collaboration time, and e-mail participants about the next steps based on the feedback. For example, if the instructional practice of using Graphic Organizers across content areas is being discussed at a staff meeting, math coaches may ask teachers to try a K-W-C chart with their students with story problems (Hyde, 2006). Before working on a problem-solving task, the teacher asks students to determine what they KNOW for sure about the task, what they WANT to find out, and to decide if there are any special CONDITIONS in this situation. The students are given time to activate their prior knowledge and make a math-to-self connection. *What does the situation in this task remind me of?*

In the example below, third graders are working on a combinations task. They have four shirts and three pants and need to determine how many combinations are possible using an organized list or other efficient strategy. A K-W-C chart would look like this:

What do you **know** for sure?	What do you **want** to find out?	Are there any special **conditions**?
Max has four shirts and three pants.	How many days can he look different?	He can only wear each thing together one time.
Show how you solved the problem using pictures, numbers and words:		
I said he had red, yellow, blue, and green shirts. Then I said he had white, black, and brown pants.	Here's what I did: r-w, r-b, r-br, b-w, b-b, b-br, y-w, y-b, y-br, g-w, g-b, g-br.	He had 12 outfits. I know 3 x 4 is 12. I got a little confused about b for blue, black, and brown, but I think I still did it right.

Students may relate this math task to something they have seen in science, social studies, the arts—a math-to-world connection. Some students may offer a math-to-math connection related to other strands in mathematics (like multiplying 4 x 3). The teacher records the students' ideas on a Graphic Organizer similar to a K-W-L chart, and then gives the students time to work on the task, independently or with another partner. The students share their solutions, looking back to check if their answer makes sense.

The teachers' K-W-C Graphic Organizer and samples of student solutions are brought to the grade level math collaboration meeting, displayed, and discussed with math coaches facilitating. The new knowledge (using a K-W-C chart) is integrated with the previous knowledge and experiences teachers had (using a K-W-L chart for reading and science). Teachers receive timely feedback on their new practice and make plans to implement this kind of Graphic Organizer once a week for the next month. Math coaches may choose to e-mail the staff a rationale for using this kind of Graphic Organizer with story problems, the sequence of steps used in the process, and things to look for in their student work. Since the next math collaboration time might be scheduled for next month, and teachers agreed to use the organizer once a week, the e-mail serves as a friendly reminder and review of the practice.

If Setting Objectives and Providing Feedback (Appendix 4) are the instructional strategies teachers are working on, coaches and teachers may want to work on generating class "Guidelines for solving story problems" like the example on the next page. This process also invests students in the scoring/evaluation process.

Guidelines for
Solving Problems and **Showing Our Work**

Solving Problems

- Read the problem carefully.
- Try to understand what the problem is asking you to do.
- Ask questions if you don't understand.
- Decide what strategy to use (make a chart, draw a picture, etc.).
- Get things from your toolkit to help, like the base ten pieces, money pieces, tile, or pattern blocks.
- See if there's any way to make the numbers friendlier, like making $3.24 into $3.25 instead, so the calculations are easier.
- Be patient and keep trying. Sometimes you have to try different ways before you find one that works.

Showing Our Work

- Explain how you did the problem.
- Show all your steps.
- If you use a sketch, label it and show what you did to get the answer.
- You don't always have to use words, numbers, and sketches, but it's good to use at least 2 different ways to show your thinking.
- Remember to show the answer clearly.
- Be neat and organized so the other person can read your work.
- Check your answer. Do the problem two different ways to see if you got it right.

The student generated guidelines for solving story problems provide effective feedback.

Some math programs provide students with unit reflections. These are particularly helpful for setting objectives and providing feedback and can be personalized to meet the needs of students. If these are not part of the math curricula, coaches, teachers, and students can create them for each unit of study and use them to establish learning goals. See Student Reflection Sheet, on opposite page.

Utilizing Additional Times

Occasionally, lunchtimes or before or after school times are utilized for meetings. In this case, coaches may plan for healthy meals/snacks to make the time more productive. Coaches need to be considerate of union issues, other content area priorities, and member's schedules when planning for their small group work.

Coaches will need to set regular meeting times and stick to those times. Everyone is busy, and whenever one meeting gets rescheduled, it impacts several others, like a chain of dominoes. Groups meet regularly (weekly, bimonthly, or monthly) to keep focused on the big ideas, take actions, share results, receive feedback, and refine and implement changes. If groups are not meeting consistently, over an extended period of time, the momentum for implementing best practices loses its energy.

Blackline A 1.14 For use after Unit One, Session 21.

NAME _____   DATE _____

Unit One Student Reflection Sheet

Go over your scored post-assessment, and think about these goals as you do. Think about what you did well and what you could improve.

The major learning goals for this unit were to:

- Improve geometry vocabulary and your ability to describe shapes
- Understand and use the rectangular array model for multiplication and division
- Draw sketches or diagrams to go with story problems
- Determine factors of whole numbers (build all the arrays for a number)

- Know whether a number is prime or composite
- Find the area and perimeter of a rectangle
- Use strategies (the ones we explored or your own) to solve multiplication facts you don't remember
- Label and explain your drawings clearly with words and/or numbers

1 What did you do well on this assessment and in this unit?

2 What are two goals for your mathematical learning that are important to you? (What could you improve?)

a

b

3 How will you meet each goal above?

a

b

At the beginning of unit instruction, students set academic goals and review them at the end of the unit.

Teachers do not like to leave their post, and coaches will want to assure them that the group work will not adversely affect their students. Some buildings make substitutes available so that study groups can meet for a half day without worrying about class coverage, or plan special events (guest speakers, fine arts, or physical activities) for the students while the teachers are meeting. Coaches examine the opportunities for small group collaboration and may need to think outside the box to find time to meet. They are flexible and creative.

District Curriculum and Assessment Work

Math coaches are often asked to participate in district curriculum alignment and assessment work. Although this time is generally spent outside the building, it directly impacts the school site. School district groups usually have a problem-solving focus. They meet to resolve an issue related to curriculum, instruction, and assessment practices and alignment. These groups are charged with completing a task, like a curriculum map for each grade level, a scope and sequence for horizontal and vertical alignment, writing assessment prompts and rubrics to meet grade-level proficiencies, or identifying key math vocabulary words needed at each grade level to support student communication. These groups may determine what every student should know and be able to do by the end of the unit/quarter/school year, how they know if they did, what materials/curriculum resources to use to get them there, and how much time it will take.

Specific people are identified for participation in these groups based on their experiences and leadership qualifications. A facilitator is appointed to ensure the work of the group is done in a timely manner. These facilitators and group members are often math content coaches. For example, math coaches might meet with classroom teachers to design a *proficiency map* based on state standards and benchmarks that can be used to inform instructional practice, conference with parents and special assistance teams, and align grading practices among second grade classroom teachers (Appendix 11). These documents are field tested by the members of the second grade teaching community and then revised later by the school district group according to the feedback. The drafts then become final documents for use by the teachers in the school or district. These small groups can get a lot done, without involving everyone in the time-consuming discussions and development of the documents. However, if all the teachers are not included in a field test or feedback loop, these papers may collect dust on a shelf and not make a bit of difference in classroom practice.

An external consultant can be brought in for initial background information on a specific concept like curriculum mapping, authentic assessment practices, or identifying essential learning, and then local coaches or facilitators with strong communication skills can lead the rest of the way.

> *When districts were looking for classroom-based journal and performance task assessments, I often presented the idea at district or schoolwide meetings. We would examine several samples of student work using a rubric to determine not just a right answer, but a performance level. Later, grade-level teams would meet and make adjustments to the prompts I shared. I would provide guidance and feedback via e-mail and phone calls, but the ultimate decisions were school or district-based. On site, teacher leaders and math content coaches completed the work with colleagues.*
>
> *Assessment consultant reflection*

Final Thoughts

Math content coaches can choose from a variety of large and small group settings to involve teachers in the process of thinking about their teaching. Teachers who engage in reflective practice are better able to support the reflection of others, and that is why group work is so powerful. With a spotlight on content, practice, reflection, and relationships, coaches establish teacher-centered professional development that provides an authentic experience of teaching as learning.

> *My principal asked an interview question that had me stumped for a bit. She wanted to know the difference between a cooperative and a collaborative staff relationship. After much reflection, I realized that cooperative teachers plan field trips together, have lunch in the lounge, and share construction paper. Collaborative teachers share their hearts and souls. They'll show you their student work, talk to you about what went wrong, celebrate their student's growth with you, challenge your thinking, and share what they wonder about. As a veteran teacher, I knew both kinds of relationships, and cherished the collaborative ones that were essential to my professional growth.*
>
> *Math coach reflection*

7

Structures for Examining Teacher Practice

"All the world is a laboratory to the inquiring mind."
Martin H. Fischer

The school site is a natural laboratory for inquiry. Math content coaches can facilitate that inquiry in a variety of ways. In this chapter, structures for looking at student work, lesson study, action research, walk-through and peer observations, and book studies are explored as ways to examine teacher practice. In a scientific inquiry, a hypothesis or initial diagnosis in response to a phenomenon is created. This is followed by the application of deductive reasoning, used to clarify, to derive, the consequences of the hypothesis. The third stage of the inquiry is inductive. In this process, the predictions are tested against the data. Inductive reasoning begins with a specific observation and seeks to apply that truth more generally. The three kinds of reasoning used in an inquiry cycle are interdependent. This process ultimately leads to an increase in knowledge and skills.

In the school setting, data creates hypotheses about student learning. These hypotheses need to be further clarified that is, taken from the general perspective to develop specific conclusions. For example, if students are not successful in Geometry on the state assessment, several hypotheses might be drawn. *Do the students experience the Geometry unit of study prior to the state assessment? Is the unit comprehensive? Does it address two- and three-dimensional figures, transformations, and measurement concepts? Is the precise vocabulary getting in the way of student proficiency?* Once these questions have been researched and clarified, and the conjectures are tested against the data, an analysis of the math curriculum is done. Pacing charts can be created to ensure that all students receive quality Geometry instruction prior to the assessment dates, and specific vocabulary, in the context of instruction, is systematically delivered.

The state assessment is given again, and the data from the following term is used to test the results of the actions against the original hypotheses. This research and response may then be generalized to other strands in mathematics instruction.

Geometry

What we know	What we wonder
There are shapes all over the world.	Do shapes have to have equal sides?
You can add shapes together.	What are angles?
Shapes can be equal, the same size.	What is a prism?
Triangles have 3 sides but can look different.	What do some of the words mean?
There is symmetry.	How many shapes are there in the world?
Geometry is 3D.	How many sides to an angle?
Circles, squares, rectangles, and triangles are geometric shapes.	What shape has 6 sides?

Students might generate a K-W-L Chart about Geometry Unit of study. This information is used as a pre-assessment and used to guide instruction.

This inquiry increases the teachers' knowledge about the state standards, adopted curriculum, and their instructional delivery, positively impacting student learning. Informally, many classroom teachers already generate and test hypotheses about their teaching practices. These structures are intended to make this process more formal, inclusive, and collaborative.

Looking at Student Work

If the school site is implementing a new standards-based math curriculum, teachers may have had an initial introduction to the core materials during a summer workshop. During the school year, they may want to look at student work with the math coaches and other teachers at their grade level to inform their instruction. Teachers have traditionally worked in isolation, but looking closely at student work offers an opportunity for teachers to reflect on their practice, assess their student progress, and approach the art of teaching from many perspectives. These authentic student artifacts become a valuable data source that is undeniably a reflection of what the teachers do in the classroom and what the students know and can do as a result of their instruction and experience. In order to create a risk-safe environment for the teachers, math coaches generally institute a protocol for examining student work that continues to foster a culture of collaboration. These procedures make it safe for teachers to ask each other questions and ensure fairness and equality in terms of how each teacher's student samples are attended to. The teachers presenting the work have the opportunity to reflect on and describe a topic or quandary, in a meaningful context—their students. Through careful examination and dialogue with colleagues, they gain insights and perspectives on their students' thinking.

The process of really looking at student work takes at least an hour. Using protocols can help manage that precious time effectively. The protocol below describes each stage and recommends how much time should be allotted for each step (Blythe & Powell, 1999).

Looking at Student Work Protocol

- ◆ Introduction to protocol, reminder of procedures, and norms (three minutes)

- ◆ Presenting teacher explains the instructional context for the student work, describes the student work, and asks a focus question: *"By the end of the year my students should be able to... In this task, they were asked to... I was wondering about..."*(seven minutes)

- ◆ Members of the group ask clarifying questions: *"How much time did they have? What other experiences had they had up to this investigation?* (seven minutes)

- ◆ Members examine the student work (five to ten minutes)

- ◆ Members of the group provide feedback on the work: *"It looks like..."*(18 to 20 minutes)

- ◆ Reflecting on the feedback: *"I guess I need to spend more time with... but overall my students..."*(three to five minutes)

Looking at student work beyond a correct answer, teachers learn about the effectiveness of their instruction, become aware of students' mathematical development, and develop a deeper understanding of how this type of assessment can drive instructional planning.

There are other times when looking at student work is for the purpose of evaluating proficiency or grading. Beyond a multiple-choice, true-false, selected response test, teachers may want to look at the process of understanding mathematics—student communication, reasoning and proof, and student use of efficient and accurate problem-solving strategies. In the case of a performance task or short and extended journal response items, teachers may want to use a holistic rubric, or develop one that is more task-specific (Appendix 12). This rubric was developed based on the performance descriptors teachers were using from the state math standards. A rubric defines the performance criteria—how good is good enough.

Basil had a collection of 39 bouncy balls. Hyden gave him 35 more bouncy balls for his birthday. How many bouncy balls does Basil have in his collection now?

$$39 + 35 = 74$$

He has 74 bouncy balls in all.

The student work sample above illustrates a one-by-one counting strategy. The child drew 39 dots in the first circle group and then drew 35 dots to the second circle, and recounted all of them to determine "he has 74 bouncy balls in all." Although the answer is correct, the strategy is inefficient because the child is making redundant counts (counting the first group, counting the second group, and then counting them all again). This child is not attending to the tens and ones as composite groups (Wright, Martland, Stafford, & Stanger, 2004)

McKell had 435 stickers in her collection before she gave 239 stickers to Katelin. How many stickers does McKell have now?

196 stickers

$$1 + 60 + 100 + 30 + 5 =$$

$$90 + 100 + 6 = 196$$

In this work sample, a third grader correctly computes the answer to a story problem using the traditional algorithm. In addition, the student provides a sketch of an open number line to show the difference between the two numbers, using friendly groups of 60, 100, 30, and 5 to reach landmark numbers (240, 300, 400, 430, 435). This work shows a complete understanding of place value and computational fluency. Using the four-point rubric, teachers would look at the criteria: connections, concepts and skills, and communication, to determine a holistic score.

Teachers are encouraged to share these rubrics with their students before the task is given to clarify the performance criteria. Math content coaches may be asked to introduce teachers to the use of generic and holistic math rubrics and open response assessments, or review what has been developed at the school site.

In this way, the practice of looking at student work can ensure the math process and content standards are being met, and curriculum units and assessments are aligned vertically among grade levels and horizontally among grade-level teams. Coaches can help promote best practice in formative and summative classroom-based assessment and encourage teachers to use their data to drive instructional planning in a timely way. When several teachers get together to look at students' work, the feedback they contribute can positively guide students' future tasks and projects.

Using Lesson Study

Lesson study is an ongoing professional development practice in which teachers collaborate to plan, observe, and refine a lesson. Japanese teachers engage in this practice and researchers have cited this work as one of the reasons for their students' high achievement in math and science. Teachers typically work on lesson study in grade-level teams, or K-2, 3–5, 6–8, or 9–12 bands. To provide focus and direction to this work, the teachers select an overarching goal and related research question that they want to explore. This goal often comes from looking at assessment data. In Japan, teachers may pursue this focus for three to four years. Here in the U. S. teachers generally study this topic for a school year. Six stages have been identified in this process (Lewis, 2002).

Lesson Study Stages

1. Planning the Lesson

2. Teaching and Observing the Lesson

3. Reflecting and Evaluating the Lesson

4. Revising the Lesson

5. Teaching, Observing, and Evaluating the Revised Lesson

6. Sharing Results

Here is an example of the process: Based on student achievement data from the district test, a team of teachers wants to investigate how to help students record their mathematical thinking and reasoning in a problem-solving context, using notation that truly represents their ideas and not just a rote procedure. Along with the math coaches, grade 3–5 teachers study their place value/computation unit and narrow their focus to a few specific lessons from their curriculum materials. As they reflect on their planning, teachers answer these questions: *How does this lesson fit into the sequence of the unit and the curriculum as a whole? What problem-solving*

Some Strategies for Solving Big Multiplication Problems

- Build or sketch a frame and fill it in, like this:

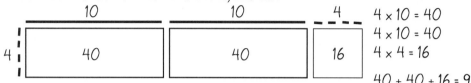

$$4 \times 10 = 40$$
$$4 \times 10 = 40$$
$$4 \times 4 = 16$$
$$40 + 40 + 16 = 96$$

- Use base 10 linear and area pieces to make an array or picture them in your head.

- Use strategies like double-doubles if you're multiplying by 4, or doubles plus one if you're multiplying by 3.

- Add the number over and over.

- Skip count, except it's kind of hard to do that if the number is very big.

- Change the number to something more friendly. 4×24 is almost like 4×25, and that's easy. It's 100. Then just take 4 away because it's really only 4×24.

In the example above, the teacher summarizes and takes notes on the student-generated strategies for solving for 4 x 24. Some students choose to use base ten and linear pieces to outline the dimensions of an array, some visualized the model/groups, others used repeated addition, skip counting, and thinking about money. As students presented their thinking, observers scribed what they saw.

strategies could be explored in this lesson? How could the teacher model and students practice a variety of notation methods that would show a range of problem-solving strategies in this lesson? In addition to reading the unit overview, coaches and teachers may also read articles and chapters in books about problem-solving strategies and student notation (problem they have selected). During this inquiry, the math coaches promote professional dialogue among the teachers to develop the lesson plan in the text more comprehensively.

One teacher from the team teaches the lesson in the classroom, and the other participants observe and take copious notes about what the teacher and students are saying and doing. In this example, they are looking for specific evidence of student reasoning and their notation practices. The data collectors are positioned in different parts of the classroom to see the class from a variety of perspectives. They experience the lesson as a student would. They do not interfere in the lesson in any way! That's hard for most teachers because they want to help struggling students. In some situations, the math coach may be teaching the lesson.

The purpose of the observation is to gather data, not to evaluate the teacher. After the lesson, the team meets to discuss the lesson and their notes. The coaches generally facilitate the conversations, while a recorder in the group takes notes. The teacher who presented the lesson speaks first about how she thinks the lesson went and identifies any problems she observed. Other team members contribute their findings, with a focus on what the students were doing and saying. On the basis of any problems that occurred, the lesson study team makes additional changes to the lesson plan and prepares for a second implementation.

This lesson is presented to a different group of students by the same teacher or another team member. Again, observers record what is happening during the actual lesson and meet to discuss their notes afterwards. Finally, the team shares their results about the topic they initially considered. *How did the changes made to the lesson impact student learning? Were the students more successful showing their problem-solving strategies using a range of notation skills? What other decisions would make the lesson more successful in the future?* The teachers' reflections and summary of group discussions document the investigation, a collaboration of teaching and learning. Lesson study groups are most effective when adults are part of a risk-safe learning community. Math content coaches can assist in that process.

> *I had no idea that this lesson was so important. In years past, I taught it as a money game and never really thought about why it was part of this unit. Now, I see what my students need to know and be able to do in order to make a connection to the other lessons in this unit on place value, fractions, and decimals. Teaching it again gave me a chance to see the difference it makes when we are intentional about our teaching practices, especially the questions we ask students.*
>
> *Teacher's reflection*

Using Action Research

In an action research project, educators work on problems they have identified for themselves. They are encouraged to examine and assess their own work, and then consider ways of working differently. Using research techniques, teachers examine their own practice systematically and carefully. This includes generating a hypothesis and collecting evidence, followed by scrupulous reflection. For example, if ability grouping in mathematics has not raised basic and proficient student achievement, teachers may look at differentiated instruction in a more student-centered way. They may delve into the research, interpret the school data, and regroup students for an upcoming unit of study. During that unit, teachers may differentiate instruction using flexible grouping, make changes in their presentations, and modify the classroom assignments to best meet the needs of the learners. Math content coaches can help organize the teaching materials, classroom environment, and assignments to promote more open cooperative groups.

Throughout the unit, the coaches might help the teachers collect formative and summative achievement data, focused especially on the data of basic and proficient students. The purpose of an action research project is to make teachers aware of what they are doing and what happens when they change what they are doing. Action research is about changing situations, not just interpreting them.

The Five Phases of Action Research

1. Problem Identification
2. Plan of Action
3. Data Collection
4. Analysis of Data
5. Plan for Future Action

Teachers identify a problem and pose it in the form of a question that is broad enough to allow for a range of insights and conclusions. Teachers decide on what practices they want to study, or change, and a timeline for doing so. Then they decide what evidence they will collect in order to answer their question. The key here is to ensure that a variety of data sources are used. If teachers had a hunch about the impact of students' disposition to math in relationship to their academic success, they would use standardized test scores, anecdotal comments from last year's teacher, portfolios of student work samples, and journal entries from this year to draw conclusions about how students feel about their performance as mathematicians. *Is there a correlation between their personal feelings and their academic achievement? Are the students who are "good in math" good at problem solving? Or just fast at answering computation tasks? What message are we sending to the students about their ability to do mathematics when we continue to emphasize timed tests?*

When the evidence is gathered, teachers and math coaches reflect on the data to verify any patterns and relationships. These insights lead to a new understanding of the impact of their practice. In the final phase, teachers decide what changes they will make, what changes they can recommend to others, and occasionally even publish their findings. Some teachers might do an action research project in their own classroom aimed at personal change while others will prefer to work in a small group setting, with coaches, focused on impacting school or grade-level changes. Since the action research project is generally teacher initiated, it's important for coaches to ask the teachers what they need and want from the inquiry.

Examining Cognitive Demand

Most teachers are attentive to the content they are required to teach in the classroom. A closer look at data sources might tell us something about what students are being asked to do with that content. For example, a classroom of fifth grade students may practice the traditional long-division algorithm and are proficient on classroom drill sheets. They are not proficient on the district and state exams when they are asked to apply an understanding of number operations by choosing the operation in the context of a problem-solving situation. Teachers may say, *"But I taught it, and they knew how to do it last week!"* In this case, there may be a misalignment in cognitive demand (Norm Webb, 1999). Asking students to do "it" when told what to do and later to apply "it" to another situation may be the root cause of the achievement discrepancy. Taking this computation skill from a recall of rote procedures and applying the learning to a higher level—a novel setting—is more cognitively complex and demanding. On the basis of this evidence, coaches and principals will want to help teachers align their practices with the cognitive demand implied by the standards and benchmarks. *What do the verbs in the NCTM Standards say students should know and be able to do? Follow procedures? Or apply and adapt a variety of appropriate strategies to solve problems?* This implementation gap can be narrowed by moving teachers forward with their mathematics curriculum adoption and a deeper analysis of the pedagogy used in the classrooms. *Are students told what and how to do mathematics or invited to solve problems using a variety of approaches? Have they had experiences talking and writing mathematics during classroom instruction?* Very few students will be able to make the cognitive shift required to be successful on constructed and extended response items without classroom practice.

Using Video Clips for Staff Development

A video clip of a standards-based lesson, where teachers and students are interacting as a community of learners is helpful for observing best practice in mathematics teaching. Videos can be home-made or professionally produced, marketed by NCTM and other publishers of math content resources. Coaches invite participants to take notes on what they see the teacher and students doing during this viewing (Appendices 6 and 7). These observations can be used to compare and contrast what the teacher's and students' role was in the lesson, and how that practice aligns to the NCTM Process Standards (Appendix 1). *How were students using problem solving, communication, and reasoning and proof to make connections and reflections during the lesson? What was the teacher's role in this process?* Again, the coaches encourage teachers to share their observations with others and help build bridges to their classroom practice.

These observation structures continue to keep participants involved in the thinking and learning, even in the context of a group setting. During the debriefing, math coaches listen actively without judgment. They keep the discussion points positive and focused on what's good for students and meaningful for teachers. Another person scribes the events of the in-service as a record of what transpired, who contributed to the discussion, and questions for further exploration.

Walk-through Observations

Scheduling an hour or more for a classroom visit is difficult sometimes. There are days when coaches or principals, or classroom teachers, have a few minutes (10 to 15 minutes) to spare. Brief classroom walk-throughs won't capture every aspect of the teaching and learning cycle, but over time, these visits and the data collected from them are noteworthy (Appendixes 6 and 7). Observation forms for walk-through visits are indispensable because time is an issue, and the recorders must stay focused on what's essential. These observations can occur more frequently and provide specific feedback on focused topics, like student engagement.

Teachers may ask coaches and principals to diagram the classroom and chart such things as the location of the students who are contributing, questioning, and doing most of the talking. This information complements data collected from other observations to give teachers feedback on the classroom dynamics that impact student involvement. Learning a variety of data collection techniques is necessary for maintaining a focus on what the teachers need and want during the observation visits. Regardless of what observation style teachers, principals, and coaches prefer, the goal of the visits is to collect accurate and complete data. They may choose a pre-made observation form to start with, and then construct one with colleagues based on their specific professional targets. This data will help the coaches plan the next steps—the next inquiry.

Planning for Peer Observations

Peer observations build collegiality and generally assist teachers in the improvement of a particular craft, usually after an initial staff development experience. When theory and demonstration are followed by guided practice and specific feedback, more teachers successfully transfer the new skill into practice (Joyce & Showers, 2002). Math content coaches may choose to encourage this partnership as another means of strengthening their one-to-one relationship. They can help teachers practice new skills, using established protocols for the pre-conference, observation, and post-conference. In some cases teachers might hear the same message a different way, and change their practice because of this peer interaction. Teachers might also problem solve the use of technology, analyze and practice an aspect of teaching, share classroom management techniques, or act as another set of eyes for special needs students or gender equity issues. This reciprocal relationship is focused on content instruction and student learning and never used for contractual summative evaluations.

I like to see other people's style, classroom management tips, and questioning strategies. I guess that's where I am right now in my own learning stage. I dislike leaving my classroom to go do anything else, but the one thing that has helped me grow as a professional is peer observations. When other teachers come to my room, they see so many things about my students and classroom that I miss when I am teaching. It's also good to know teachers struggle with the same issues I do.

Classroom teacher reflection

Getting Book Studies Started

There are many wonderful books that promote best practices in teaching and increasing student achievement in mathematics. John Van de Walle's *Elementary and Middle School Mathematics*, Cathy Fosnot's *Young Mathematicians at Work* Series, or Arthur Hyde's *Comprehending Math* are simultaneously focused on mathematics content and pedagogy. In a book study group, new and experienced teachers, at various grade levels, read the same book throughout the year or term and meet to discuss the big ideas on a regular basis. In some cases, these books are purchased for each teacher by the school, by the district, or through grant monies. These books may remain the property of the teachers after the class is completed, or be turned in to a resource library. Three major advantages of participating in a book study are the acquisition of a common math language, a deeper understanding of teaching and student learning in mathematics, and the supervised implementation of new teaching and learning strategies in mathematics. When participants also get timely feedback from their math content coaches, their professional practice may shift and serve as a springboard for continuous professional growth. As teachers reflect on their teaching, challenge each other, and construct and share materials, they positively affect student learning. When they see that effect on students, they commit to changing their practice.

The ideal size of these groups might be three to eight people for a novice coach or up to 25 participants for a veteran coach. Each group member or small group team can facilitate the discussion on a particular chapter of the book, or the math coaches may facilitate each meeting time. A written prompt may be used as a reflection or starting point for the group discussions. These prompts are most effective when they are tethered to a teacher's classroom experiences with students. For example, coaches may ask how teachers promoted discourse in the classroom this week. Many book study members can also earn recertification or graduate level credit for their work, especially if this group meets after contract hours. Districts may have a professional development plan in place with local community colleges and universities that allows math coaches to set up this credit choice. Recertification and graduate credit options can be a financial benefit for teachers who want to move along the salary schedule.

Final Thoughts

These structures help teachers examine their practice using inquiry, a variety of data collection techniques, and professional resources. Looking at student work, conducting a lesson study, engaging in action research and participating in walk-through and peer observations increase teachers' math content knowledge and improve their pedagogy, thereby increasing student achievement. Math coaches may participate in and facilitate these groups in their building, but not all in one year. Their role and responsibility is to narrow the focus and help chose the right structure and the right topic for the school site based on the goals established by the administration and teachers. Doing too much, too fast, will create chaos and fatigue and shut down the inquiry.

8

Conquering Challenges, Evaluating and Celebrating Success

*"What lies behind us and what lies ahead of us are insignificant
compared to what lies within us."*

Oliver Wendell Holmes

It's easy to get caught up in the cause and forget that leadership is a personal activity. It challenges coaches intellectually, emotionally, spiritually, and physically. With the adrenaline pumping, coaches can work themselves into believing they are indestructible. They might tune into the teachers and principals' needs, hopes, and frustrations. Their desire to fulfill the needs of others can become their demise, as coaches get caught up in the action and energy of the groups they serve and lose their internal compass. Connecting to personal needs and wisdom increases the coaches' capacity to lead. In the ongoing improvisation of content coaching, coaches act, assess, act again, reassess, and intervene again. They put themselves in the plan book and think about what they can do, can't do, and what can wait. Each non-action and action requires reflection. This "think and feel time" is critical for the coaches and also for the sustained focus of the professional learning community. Coaches continually embrace what lies within.

The Challenge of Empowering Others

The power of content coaching is in empowering others. Fostering dependency entraps people. Coaches with an exaggerated need to be needed scan the horizon for problem-solving situations, offering to fix all that is broken or looks broken. What impels them to serve people is their need to matter. As people need them more, they feel more powerful, but this sense of importance isolates them from reality. No one person can be responsible for increasing student achievement.

When coaches lose their ability to doubt, they might only pay attention to what affirms their own competence. They may forge ahead into new terrain, urging others to follow, even when they don't know where they're going. Everyone wants to have some measure of control over his or her life. Yet some people, perhaps as a product of their upbringing, have a disproportionate need for control. Their mastery at taming chaos reflects a deeper need for order. That need, and that mastery, can turn into a source of vulnerability. Consider what can happen:

> *Teachers were struggling with issues and experiencing high levels of disequilibrium; lots of chaos and conflict. Our math coach jumped in, ready and willing (and desperately needing) to take charge of the situation. Indeed, she appeared to be a hero while serving teachers in the school. Sure enough, she restored order. But her hunger for control focused her efforts on maintaining that order as an end unto itself, rather than returning the work to the whole group. Containing conflict and imposing her sense of order created a foundation for progress, but we weren't able to build on it. She became the authority and was not willing to turn over decision making to our staff. It was ultimately her vision and not ours.*
>
> *Classroom teacher reflection*

Empowerment helps teachers become expert (powerful) in their own classrooms. It is a process that develops a capacity to implement strategies and content by acting on issues they define as important. Empowerment is multidimensional. It occurs at various levels, as individuals, small groups, and whole staff communities come together. Empowerment is a social process, a journey that unfolds as teachers work through it. This process thrives on mutual respect between teachers, coaches, and building leadership. Empowerment celebrates everyone's values and beliefs and encourages the expression of issues as the shareholders define them.

For example, a capable teacher, new to fifth grade, scans the math curriculum over the summer months. He realizes that he hasn't taught multiplication and division of fractions for several years and wants to teach it differently than he did when he taught sixth grade. He approaches the math coach and together they plan a coaching cycle around the lessons in the unit that introduces fraction concepts. This coaching cycle includes studying the NCTM Focal Points for grades five and six, learning about the landmarks in understanding fraction concepts, examining his student assessment data relating to fraction skills. The math coach and teacher prepare lessons to model, team-teach, and observe one another. They make time to debrief about the instructional strategies used during the lesson and evidence of student misunderstanding and mastery.

The fifth grade team also meets for a three-hour session with the math coach to explore other models and methods for teaching and understanding multiplication and division of fractions. They agree to try and implement some additional manipulatives and visual models to support student learning in the context of the lessons in the next unit.

$$\frac{1}{4} \times 12 \quad \text{means} \quad \underline{\text{a fourth part of 12}} \text{ (objects),}$$

which is $12 \div 4 = 3$.

$$12 \times \frac{1}{4} \quad \text{means} \quad \underline{\text{12 times 1/4,}}$$

which is 3 wholes.

In other words, fourth part of 12 is the same as 12 fourth parts.

$$\frac{1}{3} \times 5 \quad \text{means} \quad \underline{\text{a third part of 5}} \text{ (objects),}$$

which is $5 \div 3 = 1\ 2/3$.

$$5 \times \frac{1}{3} \quad \text{means} \quad \underline{\text{5 times 1/3,}}$$

which is $5/3 = 1\ 2/3$.

In other words, _____ part of 5 is the same as ___ _____ parts.

In this example, a visual model helps explain the relationship between multiplication and division and division as a faction.

The math coach steps back, allowing the team to make decisions about their instruction, therefore developing capacity. The new teacher and the grade level team are empowered to make changes in their practice through professional development and collegial discussions. While math coaches cannot give teachers power and cannot make them "empowered," they continually provide the opportunities, resources, and support they need to become decision makers themselves. Math coaches ask lots of questions, challenge teachers to go beyond limited beliefs and previous patterns, celebrate with them when they achieve success, and keep the light on the ultimate vision of what is possible.

Teachers define coaches by their roles and naturally reflect their own needs and worries onto the coach. By returning the responsibility to the teachers, coaches are creating a greater capacity and giving them the greatest gift. They are consciously identifying and empowering other advocates within the community, which builds leadership density and supports sustainability.

The Challenge of Using Manipulatives

Manipulatives are objects that engage the student visual and tactile senses. They are physical models that can be used to demonstrate a mathematical idea or tools for problem solving.

Many math educators have used manipulatives in the classroom. Commercially sold manipulatives (like cubes, geoboards, pattern blocks), and home-gathered collections (like buttons, sea shells, nuts and bolts), along with computer site access, give teachers and their students many more options. Math content coaches can facilitate discussions around the following questions: *Which manipulatives are best in this particular lesson/unit? How long do we use them? Who uses them? How will these help students?*

Some teachers may be reluctant to use manipulatives because they are not familiar with how to use them to model mathematical concepts they hadn't experienced them in their own learning, they had classroom management issues during lessons that featured manipulatives, and/or they don't think they are "really doing math when the students are just building with blocks." These beliefs may interfere with an implementation of the selected standards-based curriculum that includes the use of manipulatives, but more importantly, impacts students' understanding of mathematics.

Math coaches can demonstrate a variety of manipulatives during professional development sessions and ask for other teacher leaders to model how these objects facilitate math learning. While doing so, math coaches highlight the important role teachers play in helping children make connections between the models and rich mathematics. Questions that encourage students to reflect on their actions with the materials, specifically the variables that apply to the concepts, maximize learning. Primary teachers are more likely to use manipulatives, but often when the concepts become more abstract, in the intermediate grades, the access to manipulatives is reduced. Teachers may believe that older students, or gifted students, do not need to use them. However, manipulatives have been effective in increasing achievement with elementary, middle, and high school students. They are most effective in the development of concepts prior to the introduction of algorithms (National Research Council, 2001).

In addition to the use of three-dimensional manipulatives, students need opportunities to use two-dimensional representations of those objects in the context of lessons and practice sketching their own notation. In the previous journal example, students begin to visualize the magnitude of the base ten system, without building it with pieces.

In this session, students were given base ten pieces and asked to examine the relationship amongst the pieces, using area and linear measurement dimensions.

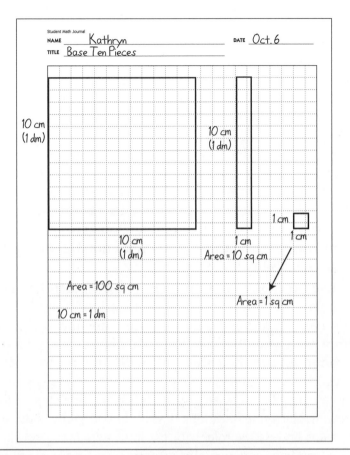

After exploring base ten manipulatives, students record the dimensions and area of each piece in their journal. They make connections to metric units centimeter, decimeter, and square centimeters using proportional relationship and measurement units.

Math coaches can facilitate effective schoolwide implementation of manipulatives by providing a rationale for their use, engaging teachers in conversations about which manipulatives would be effective, when and why, and modeling lessons that engage students in conceptual mathematics using three- and two-dimensional representations. Post-conferencing about the questioning strategies that helped students make explicit connections to the big mathematical ideas will help all teachers develop a deeper appreciation for the use of manipulatives and visual models and increase their comfort with managing materials during instruction.

The Challenge of Cooperative Group Work

Cooperative learning is an instructional paradigm in which teams of students (two to four in a group) work on structured tasks that foster positive interdependence and individual accountability through face-to-face interaction. Learners must work together to succeed. Cooperative learning improves information acquisition and retention, higher-level thinking skills, interpersonal and communication skills, and self-confidence when implemented with integrity (Johnson, Johnson, and Holubec, 2002). In competitive classroom climates, students are trying to out-do others; others must fail, so they can succeed. High-achieving students continue to "win," and low-achieving students quit because of the negative tension. The greatest benefit occurs in heterogeneous groupings, with a balance of ethnic groups, females/males, handicapped with nonhandicapped students, and students of low-to-high ability and productivity.

In order to increase motivation and self-efficacy, standards-based curricula often promote cooperative group work. Students might work in groups on assignments they would have traditionally done alone. They may collaborate on solving a problem or summarize what they have learned about ideas previously presented. They regularly communicate their thinking with one another, sharing strategies and conjectures. Using structures like a *Three-Step Interview* (Kagan), each member of a team chooses another member to be a partner. During the first step, individuals interview their partners by asking clarifying questions—what do you know about geometry? During the second step partners reverse the roles. For the final step, members share their partner's response with the team. This is especially effective for increasing engagement and language development. Math lessons may also state a group goal that requires members to play an interactive role in the completion of the task. This creates an intradependence among group members, an increase in accountability.

Students don't arrive at school knowing cooperative learning structures. They need to be taught procedures and given practice opportunities for turning and talking to a neighbor, using appropriate voice ranges, moving quietly in and out of groups, and responding to teachers' signals for attention. Math coaches can plan, model, and reflect on these cooperative learning structures with all teachers during their professional development opportunities and continue to encourage the practice—not just to increase math achievement, but also to increase self-esteem and mutual respect.

The Challenge of Working with Difficult Teachers

Don't we wish all classrooms had hard-working and caring teachers? However, in the real world coaches come across difficult teachers to work with. They are the same teachers that principals struggle to motivate, parents complain about, and students dread. Yet, they are still in the classroom. Todd Whitaker wrote about

Dealing with Difficult Teachers (2003) from a principal's point of view and offered several suggestions that math coaches need to consider.

Rules for Dealing with Difficult Teachers

♦ Keep the power in the hands of the positive teachers. Don't let thoughts about "they will not like it" get in the way of making decisions that improve instruction

♦ Deal with unprofessional behavior. Establish norms for individual and group work and hold everyone accountable for those behaviors

♦ Treat negative teachers courteously even when they seek a confrontation

♦ Catch negative teachers doing something good and praise them in public. The praise should be timely, specific, and offered without any qualifiers. For example, "I noticed Ms. R's students were using a Venn diagram to consider how the multiples of 4 and 8 compare. This strategy is one of the nine identified in the book *Classroom Instruction That Works*."

♦ At group meetings, regroup teachers, taking negative teachers out of their negative element, their comfort zone. Shuffle a deck of cards and pass them out as people arrive. Use the suits or numbers to have different teachers sit together. This shifts the balance of control and allows everyone to be more productive

♦ Maintain high expectations for the entire staff. Don't lower the bar just so everyone is successful. This increases the slackers' discomfort while moving the coaching initiative forward

♦ Try more than one strategy at a time. Be persistent. Frequent interactions make a difference

Math coaches begin with the difficult teacher's strength. *Is math or reading content knowledge a professional strength? Challenging gifted students? Working with at-risk students?* Coaches find out what the difficult teachers value. They use this information to focus on classroom teaching, assessment practices, and reflection. Each interaction is documented by the coaches using factual and objective notes, which are not evaluative or opinionated commentaries. Coaches work on influencing the negative teachers' behavior, not their personality, by introducing teaching strategies that are practiced until they become self-generating and self-correcting.

Expectations for individual teachers can be differentiated based on interest, readiness, and learning profile (Tomlinson, 1999). Not everyone has to move along at the same pace, in the same structure. However, accepting that all children can learn, treating students equitably, and incorporating best practices are not negotiable. Difficult people understand a system that is consistent. Coaches must be skilled communicators who can neutralize resistance by creating a positive climate and sense of community.

When to Facilitate, Coach, or Become the Expert

Simply put, facilitation helps a group accomplish its goals. There are a wide range of perspectives about the ideal nature and values of facilitation and leadership. For example, some coaches believe that facilitation should always be highly democratic in nature. Others believe that facilitation can be quite directive, depending on the particular stage of development of the group. Whatever one's beliefs about the best nature of facilitation, the practice is best carried out by coaches with proficient knowledge and skills regarding group dynamics and processes. Effective facilitation requires strong content knowledge and skills, also.

Teachers are directly involved in and impacted by the school's mission, and therefore they need to be part of the process that determines what they want to do and how they want to do it. As a facilitator, coaches cultivate a variety of relevant, realistic, and flexible strategies from the teachers to choose from to implement the priorities. They facilitate brainstorming sessions in small and large groups, with open-ended probing questions, and take notes on what was said. They keep in touch with the values and beliefs that teachers hold dear. Coaches analyze this information with building leadership in order to determine next steps.

Coaching, in the one-to-one setting, is a powerful means to guide and support individuals, clarify current challenges or priorities, identify instructional strategies to address those challenges, and provide modeling and follow-up to actually witness the implementation of the strategies. To maximize individual learning, coaches engage teachers in these personal experiences. Once the school year begins, the majority of the content coaches' time should be spent in this role, modeling standards-based practice, co-teaching, and observing teachers. This includes time for planning, pre- and post-conferencing.

When the initiative requires general knowledge, like a background session on content, process, or teaching standards or assessment frameworks, use of new technology systems, or guidelines for conducting general practices (lesson planning, unit planning, assessment/grading reporting, or providing tutoring services), content coaches fill in as an expert. Coaches have usually attended specialized trainings elsewhere and are asked to share the information with the teachers in the building. These trainings enhance knowledge and skills and are generally expressed in a concise and timely manner to the staff as a whole group, through demonstration and interactive sessions.

Content coaches wear all three hats: facilitator, coach, and expert, depending on the task and the participants. The challenge is to know when and where to play which role.

Reflection and Advocacy

Math coaches need to regularly reflect about teaching and learning and their coaching behaviors with principals and colleagues. Acknowledging the limits of their own competence keeps them open to learning as they blaze new trails. Coaches analyze the goals of the group for relevance, specificity, and rigor at each phase of the school's mission.

Ensuring internal and external advocacy is a priority. Advocacy is the act of arguing on behalf of an issue, idea or person. This may include trying to convince others of the importance of implementing best practices in mathematics teaching and learning. For example, coaches engage in external, standards-based assessment advocacy. They read research on best practices in assessment, share the information with others, contact school board members, and/or conduct professional development for the purpose of policy change. Internally, coaches may dispute the significance placed on drill versus practice in problem-solving settings. Again, coaches read the current research, attend conferences, put instructional ideas into practice, collect data, and essentially complete an action research project about their topic of interest. This internal advocacy is personal, life-long learning. External and internal advocacy is the search for truth and a way to communicate it effectively.

Evaluating Effort and Impact

Effort is easy to document. Coaches turn in a log of contacts, notes from pre- and post-conferences and classroom observations, and professional development trainings they have offered or attended. These attempts at making a difference mark *what* the coaches are doing. This documentation begins the first day on the job and continues through the school year. It can be as casual as a journal, or as formal as a district file.

Impact measures *how much difference* the coaches' efforts are making. For example, if schools want to measure the success of a content coaching program, they won't only be looking at the log of contacts, notes, and trainings. In addition, they look at whether teachers improved their instructional delivery and content knowledge and how well the students performed on the math assessments. Similarly, if schools want to measure the success of their math program, principals look at how well the students performed on math assessments, not at the amount of time spent in math class or the amount of professional development the teacher received. Evaluation doesn't have to be complicated, but it does require thoughtful planning.

First, coaches get information on participants' reactions to professional development experiences. Teachers let coaches know if their time was well spent; if the concepts made sense, and was the information useful (Appendix 10). Coaches also hear about the personal preferences for treats, comfort of the tables and chairs, and time for breaks and such. By attending to the personal needs of the participants, coaches increase the teacher's opportunity for attending to the learning. This feedback is generally collected by a questionnaire, and the information is used to design the delivery of future experiences.

Beyond "liking the experience," coaches want to know if the participants learned something. In this case, data is collected from a pencil-paper appraisal and used to plan the content for future professional development. Classroom coaching support for these content and pedagogy changes must also be present in order for teachers to implement them. Principals and coaches want to see teachers applying the knowledge and skills presented. The application of the concepts and skills is best observed in the classroom, several times during the year. Walk-through observations conducted by the coaches and principals will provide information regarding the effectiveness of the experience (Appendix 9).

> *An instructional strategy that we caught during the walk-throughs is the use of cues and questions.* What do you see? What do you notice? Is this familiar? How do you know? Is that always true? *These questions were followed by wait time (time right after the question was asked and after students shared to allow for the new information to sink in.) Teachers are beginning to see that these "higher-level" questions produce deeper learning than "lower-level" questions. When they wait briefly before accepting student responses, it increases the depth of students' answers. These questions are effective even before a learning experience.*
>
> *Math coach and principal reflection*

Coaches can determine the impact of their coaching on student learning via teacher implementation of best practice. This evidence, from multiple sources, is available without generating any additional student testing. Measures of student learning include portfolios, grades, and scores from standardized tests. Sometimes surveys about attitudes and dispositions toward mathematics are incorporated with data about attendance, homework completion, classroom behaviors, retention and dropout rates. These resources provide additional documentation that may help measure the impact of professional development topics.

A survey of teacher efficacy measures how teachers improved their teaching and content skills or how "effective" teachers are, based on their beliefs about their students and their teaching. Do teachers…

- ♦ View their work as meaningful and important
- ♦ Feel good about themselves and their students
- ♦ Examine students' work and their performance willingly
- ♦ Accept accountability for the outcomes/assessment data
- ♦ Believe that their students can learn
- ♦ Plan for that learning by setting goals and identifying instructional strategies to make it happen
- ♦ Involve their students in a joint venture to accomplish the learning goals

Teacher surveys are available from a variety of educational resources, including universities and commercial publishers. They generally ask teachers specific questions about their content knowledge before and after their interactions with the coaches and their knowledge of standards-based instruction strategies and the standards' impact on student learning before and after their interactions with the coaches. They also reflect on the quality of the relationship established with the coaches. School-based surveys can be developed by the building leadership team as well (Appendix 14). If teachers are going to fill out a survey, they need to know the focus and how the information will be used. *Will their name be on the survey? Will the coaches get the feedback directly or will someone pool the data together first?* Most teachers answer survey questions more truthfully if there is a privacy screen. Administration will want to examine the pre- and post-coaching survey questions for evidence of growth. According to the teachers, what specific areas were impacted by the coaching program: *Understanding of standards, teaching practices, and student learning?*

The leadership team might also consider interviewing a few teachers to explore selected issues in greater depth. The questions and dialogue should relate to the goals set at the beginning of the year and the performance indicators identified by the initial job description. For example, a survey or interview might ask selected teachers and the principal to cite specific observations and experiences which reflect on the coaches' classroom teaching skills, content knowledge, relationship development, facilitation skills, questioning skills, organization and management skills, and ongoing personal learning (Appendix 14). Coaches should self-assess their progress on the continuum. The form provided can be modified to meet the needs of the specific learning community and coaching responsibilities.

Seeking Encouragers

Coaches may provide a calm atmosphere for teachers and contain tensions during a process of organizational change. They may have the emotional and mental energy required to unite people in the midst of conflicting views and values. They may be the pillars of strength and resilience. *But who's providing that calm and emotional and mental energy for them? When they are completely exhausted, who will meet their need for intimacy and release?* Coaches will want to seek relationships with people who have a genuine interest in them and appreciate their commitment to students and teachers.

> *These people are your teachers they are your encouragers. They bless your quest and support you along the way. They cheer you on. They guide you when you lose your way, and believe me you will. They share your joy when you succeed and boost your confidence when you're disheartened. They help you evaluate options and challenge your assumptions. Most importantly, they give you the courage to act.*
>
> Kevin Carroll, *Rules of the Red Rubber Ball*

Encouragers ask the coaches hard questions like *"Why not?"* because they care about them and their work. They listen and may offer their perspective, but do not take responsibility for the decisions the coaches make. These special relationships need to be cultivated by the coaches. This inner circle of influence directs the coaches, providing positive energy and the calm during the storm.

A comprehensive collegial support program requires strong, sustained leadership to build the commitment and provide assistance for those coaches teaching and learning within the system. This includes school district administration, principals, math content coaches, coaches from other disciplines, teacher leaders, and the school community as a whole. What lies within the coaches is significant and needs to be nurtured by others whose work is focused on what's good for students and teachers.

Celebrating Success

Every organization has to determine what accomplishments are worth celebrating. Coaches celebrate little milestones with specific verbal compliments, quick notes and memos about an observed behavior, praising teachers in front of other colleagues, and organizing celebrations. Quarterly, or seasonally, they may offer to organize a celebration with other teachers or the building leadership team. The focus of these galas is to honor sincere effort and provide recognition for accomplishments that are part of the mission and to have a good time. Laughter is after all, the best medicine.

Final Comments

Many previous professional development initiatives focused on transmitting a set of routines and techniques to teachers: Put four students together during an activity, pass out manipulatives, and ask a few students to share their strategies in front of the class. These routines alone are not enough to increase student achievement. Math content coaching is about going deeper; about learning mathematics in ways that reflect a standards-based approach, about understanding complex and varied students' reasoning in mathematics, and about the power of interactions that promote discourse. Professional development via content coaching impacts student and teacher capacity and in doing so has the greatest potential of influencing classroom practice and student achievement. In order for this to transpire district and building leadership must be supportive of the coaching initiative and recognize that math coaching can only be part of a cohesive plan to increase student achievement in mathematics. Capable teachers must be willing to take on the challenge.

Expectations of what the math coaches do are clearly defined and communicated to all stakeholders. Math coaches are assigned no more than 25 teachers a year, preferably at one school or two school sites. Instructional time and staff development resources are organized to support the coaches' work with teachers. Ongoing professional development in mathematics, coaching, and facilitating adult groups is provided during the first three years, creating a sustained network for content coaches.

Math content coaching can answer the call of the National Standards (NCTM 2000, Appendix 1), Focal Points (NCTM 2006, Appendix 2), and No Child Left Behind Act (2001) by taking one step at a time toward building professional learning communities that build capacity in students, teachers, and leadership cadres.

References

Andrini, Beth (1998). *Cooperative Learning and Mathematics (K-8)*. San Clemente, CA: Kagan Publishing.

Andrini, Beth (1998). *Just a Sample (video of math lesson)*. San Clemente, CA: Kagan Publishing.

Ball, D. L. (1992). *Magical Hopes: Manipulatives and the Reform of Math Education*. Washington D.C.: American Educator 16.

Barton, M. L., & Heidema, C. (2002). *Teaching Reading in Mathematics, A Supplement to Teaching Reading in the Content Area*. Aurora, CO: Mid-continent Research for Education and Learning.

Blythe, T., Allen, D., & Powell, B. S. (1999). *Looking Together at Students' Work: A Companion to Assessing Student Learning*. N.Y. Teacher College Press.

Burns, Marilyn (2007). *About Teaching Mathematics: A K-8 Resource*, 3rd Edition. Sausalito, CA: Math Solutions Publications.

Calhoun, E. F. (1994). *How to Use Action Research in the Self-renewing School*. Alexandria, VA: Association for Supervision and Curriculum Development.

Carroll, Kevin (2004). *The Red Rubber Ball*, New York, NY: Hyperion Publishers.

Checkley, Kathy (2006). *The Essentials of Mathematics K-6: Effective Curriculum, Instruction, and Assessment (Priorities in Practice)*. Alexandria, VA: Association for Supervision and Curriculum Development.

Collins, Jim (2001). *Good to Great*. New York, NY: Harper Collins Publishers, Inc.

Covey, S. (1994). *Daily Reflections for Highly Effective People*. New York, NY: Fireside Publishing, a division of Simon and Schuster.

DuFour, R., & Eaker, R. (1998). *Professional Learning Communities at Work*. Bloomington, IN: Solution Tree.

Exemplars. No date. http:www.exemplars.com. The Web site contains a large number of performance assessment tasks and rubrics for elementary mathematics.

Fogarty, Robin & Pete, Brian (2007). *From Staff Room to Classroom*, Thousand Oaks, CA: Corwin Press.

Fosnot, C., & Dolk, M. (2001). *Young Mathematicians at Work, Number Sense, Addition and Subtraction*. Portsmouth, NH: Heinemann.

Fosnot, C., & Dolk, M. (2001). *Young Mathematicians at Work, Multiplication and Division*. Portsmouth, NH: Heinemann.

Fosnot, C., & Dolk, M. (2001). *Young Mathematicians at Work, Fractions, Decimals and Percents*. Portsmouth, NH: Heinemann.

Fullan, M. (2001). *Leading in a Culture of Change*. San Francisco, CA: Jossey-Bass.

Fullan, Michael (2003). *Leading in a Culture of Change*, presentation to the Wyoming Department of Education, School Improvement Conference, Casper, Wy.

Galvin, Kathleen M., Cooper, Pamela J. (2007) *Making Connections*, Oxford, UK: Oxford University Press.

Goleman, Daniel (2005). *Emotional Intelligence*. New York, NY: Bantam Dell, a division of Random House, Inc.

Horsley, Donald L., & Loucks-Horsley, Susan. *Journal of Staff Development*, Fall 1998 (Vol. 19, No. 4).

Hyde, Arthur. (2006). *Comprehending Math: Adapting Reading Strategies to Teach Mathematics, K-6*. Portsmouth, NH: Heinemann.

Jacobs, Heidi Hayes (2004). *Getting Results with Curriculum Mapping*. Alexandria, VA: Association for Supervision and Curriculum Development.

Johnson, David, Johnson, Roger, & Holubec, Edythe (2002). *Circles of Learning*. Alexandria, VA: Association for Supervision and Curriculum Development.

Joyce, Bruce & Showers, Beverly (2002) *Student Achievment Through Staff Development*. Alexandria, VA: Association for Supervision and Curriculum Development.

Kagan, Spencer (1994). *Cooperative Learning*. San Clemente, CA: Kagan Publishing.

Kovaleski, Joseph F. (2003). *Responsiveness-to-Intervention Symposium (RTI) Tier One, Tier Two and Tier Three Interventions*, presentation December 4–5, 2003, Kansas City, MO.

Lewis, Catherine (2002). *Lesson Study*. Northwest Regional Education Lab, Portland, OR.

Marzano, Robert J. (2003). *What Works in Schools, Translating Research into Action*. Alexandria, VA: Association for Supervision and Curriculum Development.

Marzano, R., Pickering, D., & Pollock, J. (2001). *Classroom Instruction that Works*. Alexandria, VA: Association for Supervision and Curriculum Development.

Marzano, R., Waters, T., & McNulty, B. (2005). *School Leadership that Works*. Alexandria, VA: Association for Supervision and Curriculum Development.

Math Learning Center, *Bridges in Mathematics K-5*. Salem, OR: Math Learning Center, 1999–2007.

National Council of Teachers of Mathematics (2000). *Principles and Standards for School Mathematics*. Reston, VA: NCTM.

National Council of Teachers of Mathematics (2006). *Curriculum Focal Points for K-8 Mathematics*. Reston, VA: NCTM.

National Library of Virtual Manipulatives (1999). Utah State University Web site with a wide range of manipulatives including, but not limited to, the abacus, tangrams, place value blocks, geoboards that provide interactive games and activities for practice.

National Research Council. *Adding it Up: Helping Children Learn Mathematics*. Washington, DC: National Academy Press, 2001.

Sagor, R. (1993). *How to Conduct Collaborative Action Research*. Alexandria, VA: Association for Supervision and Curriculum Development.

Schlecty, Phillip (1997). *Inventing Better Schools: An Action Plan for Educational Reform*. San Francisco, CA: Jossey-Bass.

Tomlinson, Carol Ann (1999). *Differentiated Classroom*: *Responding to the Needs of All Learners*. Alexandria, VA: Association for Supervision and Curriculum Development.

Tuckman, B. W. (1972). *Conducting Educational Research*. NY: Harcourt Brace Jovanovich.

Van de Walle, John (2007). *Elementary and Middle School Mathematics*: *Teaching Developmentally*. Pearson Publishing.

Webb, Norman L. (1999). *Alignment of Science and Mathematics Standards and Assessments in Four States*, Council of Chief State School Officers.

Whitaker, Todd (2003). *Dealing with Difficult Teachers*. Larchmont, NY: Eye on Education.

Wright, R., Martland, J., Stafford, A., & Stanger, G. (2004). *Teaching Number: Advancing Children's Skills and Strategies*. London, England: Paul Chapman Publishing, Ltd.

Zemke, Ron and Susan. 30 Things We Know For Sure About Adult Learners, *Innovation Abstracts*, Vol. VI, No. 8, March 9, 1984.

Zemelman, S., Daniels, H., and Hyde, A. (1998) *Best Practice New Standards for Teaching and Learning in America's Schools*. Portsmouth, NH: Heinemann.

Appendices

Process Standards for School Mathematics

Instructional programs from pre-kindergarten through grade 12 should enable all students to:

Problem Solving Standard

♦ Build new mathematical knowledge through problem solving;
♦ Solve problems that arise in other contexts;
♦ Apply and adapt a variety of appropriate strategies to solve problems;
♦ Monitor and reflect on the process of mathematical problem solving.

Reasoning and Proof Standard

♦ Recognize reasoning and proof as fundamental aspects of mathematics;
♦ Make and investigate mathematical conjectures;
♦ Develop and evaluate mathematical arguments and proofs;
♦ Select and use various types of reasoning and methods of proof.

Communication Standard

♦ Organize and consolidate their mathematical thinking through communication;
♦ Communicate mathematical thinking coherently and clearly to peers, teachers, and others;
♦ Analyze and evaluate mathematical thinking and strategies of others;
♦ Use the language of mathematics to express mathematical ideas precisely.

Connections Standard

♦ Recognize and use connections among mathematical ideas;
♦ Understand how mathematical ideas interconnect and build on one another to produce a coherent whole;
♦ Recognize and apply mathematics in contexts outside of mathematics.

Representation Standard

♦ Create and use representation to organize, record, and communicate mathematical ideas;
♦ Select, apply, and translate among mathematical representations to solve problems;
♦ Use representations to model and interpret physical, social, and mathematical phenomena.
♦ Content Standards for School Mathematics

Content Standards for School Mathematics

Number and Operation Standard

♦ Understand numbers, ways of representing numbers, relationships among numbers, and number systems;
♦ Understand meanings of operations and how they relate to one another;
♦ Compute fluently and make reasonable estimates.

Algebra Standard

♦ Understand patterns, relationships, and functions;
♦ Represent and analyze mathematical situations and structures using algebraic symbols;
♦ Use mathematical models to represent and understand quantitative relationships;
♦ Analyze change in various contexts.

Geometry Standard

♦ Analyze characteristics and properties of two- and three- dimensional geometric shapes and develop mathematical arguments about geometric relationships;
♦ Specify locations and describe spatial relationships using coordinate geometry and other representational systems;
♦ Apply transformations and use symmetry to analyze mathematical situations;
♦ Use visualization, spatial reasoning, and geometric modeling to solve problems.

Measurement Standard

♦ Understand measurable attributes of objects and the units, systems, and processes of measurement;
♦ Apply appropriate techniques, tools, and formulas to describe measurements.

Data Analysis and Probability Standard

♦ Formulate questions that can be addressed with data and collect, organize, and display relevant data to answer them;
♦ Select and use appropriate statistical methods to analyze data;
♦ Develop and evaluate inferences and predictions that are based on data;
♦ Understand and apply basic concepts of probability.

Curriculum Focal Points and Connections for Grade 3

The set of three curriculum focal points and related connections for mathematics in grade 3 follow. These topics are the recommended content emphases for this grade level. It is essential that these focal points be addressed in contexts that promote problem solving, reasoning, communication, making connections, and designing and analyzing representations.

Grade 3 Curriculum Focal Points

Number and Operations and Algebra: **Developing understandings of multiplication and division and strategies for basic multiplication facts and related division facts**

Students understand the meanings of multiplication and division of whole numbers through the use of representations (e.g., equal-sized groups, arrays, area models, and equal "jumps" on number lines for multiplication, and successive subtraction, partitioning, and sharing for division). They use properties of addition and multiplication (e.g., commutativity, associativity, and the distributive property) to multiply whole numbers and apply increasingly sophisticated strategies based on these properties to solve multiplication and division problems involving basic facts. By comparing a variety of solution strategies, students relate multiplication and division as inverse operations.

Number and Operations: **Developing an understanding of fractions and fraction equivalence**

Students develop an understanding of the meanings and uses of fractions to represent parts of a whole, parts of a set, or points or distances on a number line. They understand that the size of a fractional part is relative to the size of the whole, and they use fractions to represent numbers that are equal to, less than, or greater than 1. They solve problems that involve comparing and ordering fractions by using models, benchmark fractions, or common numerators or denominators. They understand and use models, including the number line, to identify equivalent fractions.

Geometry: **Describing and analyzing properties of two-dimensional shapes**

Students describe, analyze, compare, and classify two-dimensional shapes by their sides and angles and connect these attributes to definitions of shapes. Students investigate, describe, and reason about decomposing, combining, and transforming polygons to make other polygons. Through building, drawing, and analyzing two-dimensional shapes, students understand attributes and properties of two-dimensional space and the use of those attributes and properties in solving problems, including applications involving congruence and symmetry.

Connections to the Focal Points

Algebra: Understanding properties of multiplication and the relationship between multiplication and division is a part of algebra readiness that develops at grade 3. The creation and analysis of patterns and relationships involving multiplication and division should occur at this grade level. Students build a foundation for later under-standing of functional relationships by describing relationships in context with such statements as, "The number of legs is 4 times the number of chairs."

Measurement: Students in grade 3 strengthen their understanding of fractions as they confront problems in linear measurement that call for more precision than the whole unit allowed them in their work in grade 2. They develop their facility in measuring with fractional parts of linear units. Students develop measurement concepts and skills through experiences in analyzing attributes and properties of two-dimensional objects. They form an understanding of perimeter as a measurable attribute and select appro-priate units, strategies, and tools to solve problems involving perimeter.

Data Analysis: Addition, subtraction, multiplication, and division of whole numbers come into play as students construct and analyze frequency tables, bar graphs, picture graphs, and line plots and use them to solve problems.

Number and Operations: Building on their work in grade 2, students extend their understanding of place value to numbers up to 10,000 in various contexts. Students also apply this understanding to the task of representing numbers in different equiva-lent forms (e.g., expanded notation). They develop their understanding of numbers by building their facility with mental computation (addition and subtraction in special cases, such as 2,500 + 6,000 and 9,000 − 5,000), by using computational estimation, and by performing paper-and-pencil computations.

Increase Attention to:	Decrease Attention to:
Teaching Practices	**Teaching Practices**
Use of manipulative materials	Rote practice
Cooperative Group Work and the discussion of mathematics	Rote memorization of rules and formulas
Writing about mathematics	Single answers and single methods to find answers
Questioning and making conjectures	Use of drill worksheets
Justification of thinking	Repetitive written practice
Being a facilitator of learning	Teaching my telling
Problem solving approach to instruction	Teaching computation out of context
Content integration	Stressing memorization
Assessing learning as an integral part of instruction	Testing for grades only
Process Standards	**Process Standards**
Word problems with a variety of structures and solution paths	Use of cue words to determine operation to be used
Open-ended problems and extended problem-solving projects	Practicing routine, one-step problems
Investigating and formulating questions from problem situations	Practicing problems categorized by types
Discussing, reading, writing, and listening to mathematical ideas	Doing fill-in-the-blank, yes-no, numerical responses
Drawing logical conclusions, justifying answers and solution processes	Relying on authorities (teacher and answer key)
Connecting mathematics to other subjects and the real world	Learning isolated topics
Applying mathematics	Developing skills out of context

Increase Attention to:	Decrease Attention to:
Content Standards	**Content Standards**
Developing number and operation sense	Early use of symbolic notation, tedious paper and pencil computation
Thinking strategies for basic facts	
Developing spatial sense	Memorizing rules and procedures without understanding
Applying skills to problem-solving situations	Memorizing formulas
Collection and organization of data	Memorizing facts and drilling
Describe, analyze, evaluate, and make decisions about information	
Evaluation	**Evaluation**
Having assessment be an integral part of teaching	Having assessment be simply counting correct answers for grades
Focusing on a broad range of mathematical tasks and taking a holistic view	Focusing on a large number of specific and isolated skills
Developing problem situations that require applications of a number of mathematical ideas	Using exercises or word problems requiring only one or two skills
Using multiple assessment techniques, including written, oral, and demonstration formats	Using only written tests

From: *Best Practice New Standards for Teaching and Learning in America's Schools*, S. Zemelman, H. Daniels and A. Hyde (1998)

From *Classroom Instruction that Works*
Marzano, Pickering & Pollock

1. Identifying Similarities and Differences

Presenting students with explicit guidance in identifying similarities and differences enhances students' understanding of and ability to use knowledge.

Asking students to independently identify similarities and differences enhances students' understanding of and ability to use knowledge.

Representing similarities and differences in graphic or symbolic form enhances students' understanding of and ability to use knowledge.

Identification of similarities and differences can be accomplished in a variety of ways. The identification of similarities and differences is a highly robust activity. (comparing, classifying, creating metaphors, and analogies).

2. Summarizing and Note Taking

To effectively summarize, students must delete some information, substitute some information, and keep some information.

To effectively delete, substitute, and keep information, students must analyze the information at a fairly deep level.

Being aware of the explicit structure of information is an aid to summarizing information.

3. Reinforcing Effort and Providing Recognition

Not all students realize the importance of believing in effort.

Students can learn to change their beliefs to an emphasis on effort.

Rewards do not necessarily have a negative effect on intrinsic motivation.

Reward is most effective when it is contingent on the attainment of some standard of performance.

Abstract symbolic recognition is more effective than tangible rewards.

4. Homework and Practice

The amount of homework assigned to students should be different from elementary to middle school to high school.

Parent involvement in homework should be kept to a minimum.

The purpose of homework should be identified and articulated.

If homework is assigned, it should be commented on.

Mastering a skill requires a fair amount of focused practice.

While practicing, students should adapt and shape what they have learned.

5. Non-linguistic Representations

A variety of activities produce non-linguistic representations (graphic representations, making physical models, generating mental pictures, drawing pictures and pictographs, engaging in kinesthetic activity).

6. Cooperative Learning

Organizing groups based on ability levels should be done sparingly.

Cooperative groups should be kept rather small in size.

Cooperative learning should be applied consistently and systematically, but not overused.

7. Setting Objectives and Providing Feedback

Instructional goals narrow what students focus on.

Instructional goals should not be too specific.

Students should be encouraged to personalize the teacher's goals.

Feedback should be "corrective" in nature.

Feedback should be timely.

Feedback should be specific to a criterion.

Students can effectively provide some of their own feedback.

8. Generating and Testing Hypotheses

Hypotheses generation and testing can be approached in a more inductive or deductive manner.

Teachers should ask students to clearly explain their hypotheses and their conclusions.

9. Cues, Questions & Advanced Organizer

Cues and questions should focus on what is important as opposed to what is unusual.

"Higher-level" questions produce deeper learning than "lower-level" questions.

"Waiting" briefly before accepting responses from students has the effect of increasing the depth of students' answers.

Questions are effective learning tools even when asked before a learning experience.

Appendix 5

Name of teacher _____ Date _____

Name of lesson _____

_____

1. What is this unit about? What mathematics will students investigate?

2. What is this lesson about? What mathematics will students investigate? How does this align to your grade level benchmarks and standards?

3. What do students need to know and be able to do in order to be successful with this lesson? How will you know if they are?

4. What concepts might students have misconceptions or difficulties with?

5. What support or interventions will you use during this lesson?

6. What extensions or challenges will you offer?

7. What instructional strategies will you employ during the lesson? Why?

8. What specific teaching or learning focus would you like feedback on?

Classroom Look-Fors

Teacher's Name _____ Date _____

Communication	Questioning
Think-Pair-Shares are used to engage students in discussions. Clear and precise mathematical language is modeled. Teachers are talking less. Students are talking more.	Teachers ask a variety of questions including some "What is this? (one correct response) And "How?" and "Why?" (higher order thinking)
Student Engagement	**Visual Models/Manipulatives**
Students are actively *doing* mathematics. Teachers are facilitating students sharing strategies with partners or the class.	Teachers insure that students have access to a variety of models/manipulatives to support their communication and reasoning.
Monitoring Student Learning	**Community of Learners**
Teachers are aware of grade level benchmarks and are actively listening for misconceptions and mastery. Appropriate challenge and support is provided.	Projects and posters are displayed around the classroom to communicate the math culture. Student comments and observations are posted. Students respectfully listen and engage in discussions about mathematical ideas.

Additional Notes:

Appendix 7

Classroom Observation Form

Environment, Student and Teacher

Name of teacher _____ Date _____

Lesson _____

_____

Physical environment:

♦ Evidence of student work

♦ Vocabulary for the current unit displayed

♦ Classroom is set up for small group work

Students (give specific examples below each item you observe)

♦ Interacting with each other, as well as working independently, while solving complex tasks.

♦ Applying problem solving strategies to mathematical situations, using the adopted curriculum (not just practicing a collection of isolated skills).

♦ Communicating mathematical ideas to one another through demonstrations, models, drawings, and logical arguments.

♦ Helping to clarify each other's learning through discussion and modeling.

Teachers (give specific examples below each item you observe)

♦ Facilitating meaningful, using complex math problems.

♦ Moving around the room monitoring learning and asking questions to challenge and clarify student thinking.

♦ Encouraging students to consider other possibilities and use more than one strategy to solve a problem.

♦ Leading students through discussions about and understanding of the important mathematics in the investigation. Providing closure when appropriate.

Appendix 8

Name of teacher _____ Date _____

Name of Observer _____

Teacher comments: How did the lesson go? What went well? What did you struggle with?	Observer comments & questions: I noticed... I wonder about...
What will you do differently with this lesson next year, next time?	Observer questions & suggestions: What if...? Could you...?
Based on data and feedback, teacher action to be taken:	Based on data and feedback, observer/coach action to be taken:

Appendix 9

Log for Working with Staff

Date: _____

Teacher	Action/Lesson/Skill	Comments

Appendix 10 — Staff Development Feedback Form

Please share your feedback with us.

1. How did you feel about the math tasks? Was it interesting, engaging? What have you done in the past that was like this? What will you take back to the classroom from this task?

2. What instructional strategies did I use to facilitate the instruction or session? What strategies will you implement in your classroom?

4. How was this helpful to you as a learner/teacher? How will these strategies help all your students?

5. What are your individual needs for the school year? Ideas for next steps?

Appendix 11

Second Grade Proficiency Map

Note: This proficiency map is in draft form. It reflects the minimum competencies for each benchmark, for each quarter, as second graders work towards proficiency. We believe that these concepts and skills are best assessed using performance tasks, interviews, paper/pencil assessments, and observations. Our own teaching is well and above these benchmarks.

First Quarter	Second Quarter	Third Quarter	Fourth Quarter
Content Standard 1: NUMBER OPERATIONS AND CONCEPTS			
Students use the concept of place value to read and write numbers up to 250.	Students use the concept of place value to read and write numbers up to 500.	Students use the concept of place value to read and write numbers up to 750.	Students use the concept of place value to read and write numbers up to 999.
Students compare and order whole numbers to 250.	Students compare and order whole numbers to 500.	Students compare and order whole numbers to 750.	Students compare and order whole numbers to 999.
Students use coins to compare the value and make combinations up to 25 cents.	Students use coins to compare the value and make combinations up to 50 cents.	Students use coins to compare the value and make combinations up to 75 cents.	Students use coins to compare the value and make combinations up to one dollar.
Students demonstrate computational fluency with basic facts using counting on, doubles, neighbors, and making ten as strategies (adds to 12).	Students demonstrate computational fluency with basic facts using fast tens, fast nines, and leftovers as strategies (adds to 20).	Students demonstrate computational fluency with basic facts (subtraction zero facts, counting back, doubles, neighbors, and half facts strategies to subtract from ten).	Students demonstrate computational fluency with basic facts (adds to 20, subtract from 20) using a variety of efficient noncount by one strategies.
Students use mental math (fact families to ten) to solve problems.	Students use mental math (fact families to 20) to solve problems.	Students use mental math (fact families) and estimation strategies (referent to a group of ten) to solve problems.	Students use mental math (fact families) and estimation strategies (referent to a group of ten) to solve problems.
Students look for a pattern as a strategy to solve problems.	Students look for a pattern as a strategy to solve problems.	Students look for a pattern and use guess and check as strategies to solve problems.	Students look for a pattern and use guess and check as strategies to solve problems.
Developing.	Students communicate their choice of procedures and results when performing operations in a problem-solving situation in one way (pictures, numbers, words).	Students communicate their choice of procedures and results when performing operations in a problem-solving situation in two ways (pictures, numbers, words).	Students communicate their choice of appropriate grade level procedures and results when performing operations in a problem-solving situation (pictures, numbers, words).
Content Standard 2: GEOMETRY			
Developing.	Students name two- and three-dimensional objects.	Students name and describe two- and three-dimensional geometric objects.	Students name, compare, and describe two- and three-dimensional geometric objects.
Developing.	Students identify lines of symmetry in various two-dimensional geometric objects.	Students identify lines of symmetry in two- and three-dimensional geometric objects.	Students identify lines of symmetry in various geometric objects.
Developing.	Developing.	Students select and use organizational methods in a problem-solving situation using two-dimensional objects.	Students select, use, and communicate organizational methods in a problem-solving situation using two- and three-dimensional objects.

First Quarter	Second Quarter	Third Quarter	Fourth Quarter
Content Standard 3: MEASUREMENT			
Developing.	Students explore estimation and apply measurement of length to content problems using standard units, up to 50 inches.	Students explore estimation and apply measurement of length to content problems using standard units, up to 100 inches.	Students apply estimation and measurement of length to content problems using standard units to the nearest inch.
Developing.	Developing.	Students apply measurement of weight to content problems using up to 100 non-standard units.	Students apply estimation and measurement of weight to content problems using non-standard units.
Developing.	Developing.	Students tell time, using both analog and digital clocks to the nearest 15 minutes.	Students tell time, using both analog and digital clocks to the nearest five minutes.
Content Standard 4: ALGEBRAIC CONCEPTS			
Students recognize and copy growing patterns using manipulatives and graphic representations.	Students recognize and create growing patterns using manipulatives and graphic representations.	Students recognize, describe, and create growing patterns using manipulatives and graphic representations.	Students recognize, describe, create, and extend growing patterns using manipulatives and graphic representations.
Students apply knowledge of appropriate grade-level patterns when solving problems.	Students apply knowledge of appropriate grade-level patterns when solving problems.	Students apply knowledge of appropriate grade-level patterns when solving problems.	Students apply knowledge of appropriate grade-level patterns when solving problems.
Content Standard 5: DATA ANALYSIS AND PROBABILITY			
Students collect and organize data using a Venn diagram.	Students collect and organize data using graphs and Venn diagrams.	Students collect and organize data using graphs and Venn diagrams.	Students collect, organize, and report data using graphs and Venn diagrams.
Students communicate conclusions about a set of data using Venn diagrams.	Students communicate conclusions about a set of data using graphs and Venn diagrams.	Students communicate conclusions about a set of data using graphs and Venn diagrams.	Students communicate conclusions about a set of data using graphs and Venn diagrams.
Developing.	Students explore simple probability experiments using spinners.	Students perform and record results of simple probability experiments using equally and unequally divided spinners with minimal errors.	Students perform and record results of simple probability experiments using equally and unequally divided spinners

Appendix 12   **Holistic Mathematics Rubric**

Criteria	4 Advanced	3 Proficient	2 Basic	1 Below Basic
Connections	Makes *complex* connections among mathematical ideas.	Makes *relevant* connections among mathematical ideas.	Makes *simple* connections among ideas.	Makes no connections.
Concepts and Skills	Demonstrates an understanding of concepts and skills *above* grade level.	Demonstrates an understanding of concepts and skills *at grade* level.	Demonstrates an understanding of concepts and skills *below* grade level.	Little or no evidence of concepts and skills.
Language/ Communication	Uses clear and coherent mathematical language to justify reasoning.	Uses mathematical language to communicate sound reasoning.	Uses minimal or incorrect mathematical language to communicate thinking.	Needs extensive support to communicate.

Appendix 13

Reflections on Mathematics Teaching and Learning

1. Do I encourage my students to remember a process or method or do I encourage them to look at problems in their own way?

2. Do I encourage my students to take part in hands-on explorations with manipulatives or do I limit their use to demonstrations?

3. Do I encourage students to share their strategies and help one another?

4. Do I encourage students to rely on their own thinking or am I most always the authority?

5. When students are working do I circulate to observe, ask questions, and try to discover students' thinking and understanding?

6. Do I provide work choices for my students?

Appendix 14

Evaluating Coaching Progress

Performance Categories	Advanced	Proficient	Basic
Classroom teaching skills: (seen during modeling, teaching observations)	Coach is accomplished at teaching numerous grade levels. Coach implements a variety of instructional strategies to increase student engagement and achievement.	Coach is capable of teaching several grade levels and pursuing more experiences. Coach implements some instructional strategies to increase student engagement.	Coach is able to teach in a narrow grade level range. Little or no evidence of instructional strategies that define best practice.
Content knowledge: (seen during conferences, small and whole group professional development)	Coach has a rich understanding of the math content and how it is organized, developed, linked to other disciplines, and applied to real-world settings.	Coach has an adequate understanding of the math content and is pursuing additional knowledge and skills.	Coach has a basic understanding of the math content, and applies rules and procedures to find the answers.
Relationship development: (seen during formal and informal interactions)	Coach employs a wide range of interpersonal strategies to build trusting and respectful relationships with all teachers. This promotes deeper reflection and classroom-based inquiry.	Coach builds relationships with a range of teachers, including some difficult people. This promotes conversations on best practices.	Coach builds relationships with like-minded teachers. This promotes conversations about what they are doing right.
Facilitation skills: (seen during small and whole group gatherings)	Coach is accomplished at facilitating small and large groups. The coach implements a variety of facilitation styles to develop teacher efficacy.	Coach is capable of facilitating small and large group gatherings. The coach implements some strategies to manage groups and get things done.	Coach is able to facilitate small group gatherings by creating a consistent structure.
Questioning skills: (seen in post-conferences after classroom observation)	Coach encourages teachers to reflect on their practice by asking open-ended, probing questions and examining student work. Coach also reflects on the art of teaching.	Coach asks questions to elicit reflection about student learning and sometimes teacher practice. Coach keeps busy and doesn't take time for reflection.	Coach asks closed questions about student learning and makes no connection to teaching practice.
Organization and management skills: (seen in contact logs and teacher surveys)	Coach uses data to drive decisions about coaching one-on-one, or working with small or whole groups. Coach has developed a consistent schedule that impacts student learning and teacher effectiveness.	Coach makes decisions about coaching in response to teacher or principal request. Coach has a schedule that generally works.	Coach makes decisions about coaching in reaction to a crisis. Coach has not established a schedule and thereby is unable to follow through on agreements.
Ongoing personal learning: (seen in contact logs and informal and formal contacts)	Coach engages in a variety of professional development to optimize personal learning. Coach seeks to balance the current research and their personal classroom experience.	Coach attends professional development to learn more about mathematics content and pedagogy.	Coach attends mandatory professional development.

Notes

Notes